Fire Prevention in Modern Construction

Fires continue to be an increasing threat to buildings and construction sites. Under construction, buildings are very vulnerable to fire hazards. Modifications made to buildings over time, whether for maintenance, repair or improvements, can significantly hinder a building's life safety degree if fire safety is not considered.

Fire Prevention in Modern Construction: A Practical Guide introduces and explains the fundamentals of construction fire safety. It explores applicable codes and standards while providing a fundamental understanding of common fire protection and life safety systems found in buildings. It offers guidance on inspection techniques and how to foster cooperative communication between the Fire Protection Program Manager (FPPM), design team, general contractor, subcontractors, authority having jurisdiction (AHJ), and owners. The book features several examples and real-life applications, highlighting best practices that can be applied to a variety of job sites.

Written by a fire protection engineer with extensive experience developing, implementing, and enforcing construction fire safety plans, this book is an essential resource for those serving as a FPPM, but also equips professionals in the construction, real estate, engineering, and property management sectors with the necessary fire prevention fundamentals to support real estate projects and keep them safe.

Matthew J. Bourque is a fire protection engineer who is currently the Director of Fire Protection and Construction Operations at a leading real estate development firm. Matthew is involved in the Boston Seaport development, the largest single real estate project in city history. He graduated from Worcester Polytechnic Institute with a Bachelor of Science in Mechanical Engineering and a Master of Science in Fire Protection Engineering. He is a registered professional engineer (P.E.) in Fire Protection Engineering and is also a certified local building inspector and licensed construction supervisor. He was appointed to serve on the Committee on Construction and Demolition for NFPA 241, Standard for Safeguarding Construction, Alteration, and Demolition Operations and was selected to speak at the National Fire Protection Association (NFPA) headquarters on construction fire safety as well as support the development of training material related to fire protection program management. In partnership with the NFPA, he created a training course for the Fire Protection Program Manager role. His background and construction experience provide him with a unique perspective on construction fire safety.

"I have worked closely with fire protection professionals throughout my career on many large development projects. This book provides fundamental fire prevention concepts, invaluable information on fire protection systems, and best practices that are critical for managing fire risks effectively."

– **John "Jack" Dempsey**, *Ret. Fire Commissioner, Boston Fire Department*

Fire Prevention in Modern Construction

A Practical Guide

Matthew J. Bourque

Routledge
Taylor & Francis Group
NEW YORK AND LONDON

Designed cover image: Shutterstock

First published 2026
by Routledge
605 Third Avenue, New York, NY 10158

and by Routledge
4 Park Square, Milton Park, Abingdon, Oxon, OX14 4RN

Routledge is an imprint of the Taylor & Francis Group, an informa business

Library of Congress Cataloging-in-Publication Data
Names: Bourque, Matthew J. author
Title: Fire prevention in modern construction : a practical guide / Matthew J. Bourque.
Description: New York, NY : Routledge, 2026. | Includes bibliographical references.
Identifiers: LCCN 2025048433 (print) | LCCN 2025048434 (ebook) | ISBN 9781032720203 hardback | ISBN 9781032698533 paperback | ISBN 9781032720210 ebook
Subjects: LCSH: Building, Fireproof | Construction industry--Fires and fire prevention
Classification: LCC TH1065 .B68 2026 (print) | LCC TH1065 (ebook)
LC record available at https://lccn.loc.gov/2025048433
LC ebook record available at https://lccn.loc.gov/2025048434

ISBN: 978-1-032-72020-3 (hbk)
ISBN: 978-1-032-69853-3 (pbk)
ISBN: 978-1-032-72021-0 (ebk)

DOI: 10.1201/9781032720210

Typeset in Times New Roman
by KnowledgeWorks Global Ltd.

To my children, whose love, curiosity, and laughter inspire me each day.

Contents

About the Author

Matthew J. Bourque is a fire protection engineer with extensive experience in construction fire prevention. He is the Director of Fire Protection and Construction Operations at a leading real estate development firm. Matthew is involved in the Boston Seaport development, the largest single real estate project in city history. The project includes the transformation of 33 acres of land and 7.6 million square feet of mixed-use buildings, including retail, office, lab, entertainment and residential uses. He oversees the development of the construction fire safety program and the implementation of the plan across various projects. This includes the close coordination with the Boston Fire Department (BFD) and Inspectional Services Department (ISD), among other fire and life safety-related duties. He ensures continuity of the construction fire safety program for all tenant improvement projects across the district. He oversees firestopping and other life safety special inspections. Matthew manages the inspection and testing of fire protection systems and the overall maintenance of the systems. Matthew leads construction fire safety-focused meetings with contractors, design team members, authorities having jurisdiction (AHJ) and insurance representatives to ensure projects are meeting construction fire safety standards.

His experience also expands into construction management as he has led multiple redevelopment projects that have touched many jurisdictions across the United States. Matthew has led efforts to develop and ultimately receive approval on construction fire safety and phasing plans in various municipalities.

Matthew's commitment to fire protection engineering has also led him to start a fire protection engineering consulting company. He works closely with owners, developers, architects and engineers to develop fire protection strategies in a wide array of building types and uses. He develops engineered solutions tailored for the project's needs. This includes the design of fire protection systems and navigating building and fire codes.

Previously, Matthew worked as a building and fire code consultant for a fire protection engineering company. He graduated from Worcester Polytechnic Institute with a Bachelor of Science in Mechanical Engineering and a Master of Science in Fire Protection Engineering. Matthew is a registered professional engineer (P.E.) in Fire Protection Engineering. He is also a certified local building inspector and licensed construction supervisor. Matthew has been appointed to serve on the Committee on Construction and Demolition for NFPA 241, *Standard for Safeguarding Construction, Alteration, and Demolition Operations.* Matthew was selected to speak at the National Fire Protection Association (NFPA) headquarters on construction fire safety, as well as support the development of training material related to fire protection program management. He also participates in the Construction Fire Safety Partnership, an organization focused on improving construction fire safety. He also serves on the Technical Advisory Committee for the Massachusetts Board of Building Regulations and Standards (BBRS), a state-appointed board for managing Massachusetts's building codes and construction licensing. Matthew's background and construction experience provide him with a unique perspective on construction fire safety.

Matthew brings extensive experience in developing, implementing, and enforcing construction fire safety plans across a wide range of project types. His perspective is particularly valuable, as most professionals in the field are typically involved in only one aspect of the process. Engineers who create the plans lack first-hand experience in the implementation of them by not fully grasping the realities of a construction site. Additionally, contractors who implement the plan may not have the foundational knowledge of fire protection engineering principles and codes and regulations that influence the plan.

Preface

Fires continue to be an increasing threat to buildings and construction sites. The frequency of news headlines associated with construction-related fires validates the importance of prioritizing construction fire safety. Modifications made to buildings over time, whether for maintenance, repair or improvements, can significantly hinder a building's life safety degree if fire safety is not considered. The Fire Protection Program Manager's role in the process is critical to ensure construction sites and operating buildings are maintained at an acceptable level of safety.

The book is structured into three different parts:

Part I: Introduction to Fire Prevention in Buildings & Construction Fire Safety
Part II: Fire Protection Systems and Life Safety Features
Part III: Best Practices and Applications

Part I is intended to bring to light the need for fire prevention considerations in buildings. A brief history of fire and building codes and their more recent developments will be highlighted. Influencing codes and standards as they relate to building fire protection will be explored. The Construction Fire Safety Plan (CFSP) and the critical role of the Fire Protection Program Manager (FPPM) will be introduced.

Part II is intended to give readers a basic understanding of common fire protection and life safety systems found in buildings. This foundational material is presented so FPPMs, contractors, owners and property managers alike can satisfactorily manage these systems. The information is outlined in a manner that allows easy navigation to bring value to those who may be in a full-time fire prevention role, or to those with minimal involvement who just need an overall understanding to successfully carry out their job function.

Part III provides recommended best practices and real-life applications. Examples, templates and checklists are provided throughout this section to allow readers to implement proven and effective methods in their own projects. This section covers common construction site hazards. It also touches on critical collaboration between the project team and fire departments. It explores the transition to an occupiable building.

The motivation behind this book is driven by the limited comprehensive resources on construction fire prevention and overall building fire safety. While codes, insurance requirements and fire department regulations have increasingly emphasized fire prevention in buildings, there remains a gap in practical guidance. As more and more people in the construction and real estate industries enter into Fire Protection Program Manager (FPPM) or other fire prevention roles, the expectations are not clearly defined. The qualifications, including education, knowledge and experience, have been ambiguous. Processes, inspections and protocols have not been clearly defined. This book is intended to help close that gap for those serving as Fire Protection Program Managers, as well as any person directly or indirectly involved in construction projects support the construction fire safety program and ultimately prevent fires.

Foreword

As the 1970 founder of Cummings Properties in Woburn, Massachusetts, I have believed for many decades that the "success" of any real property development goes far beyond financial gains. We are strong believers in the importance of contributing positively to society. My wife Joyce and I started Cummings Foundation, Inc. in 1986, as a mechanism to give back. The foundation continually awards over $50 million each year to a diverse group of nonprofits, primarily in Eastern Massachusetts, and its total actual grants to date exceed $600 million.

With well more than half a century of experience in the real estate industry, and having built a portfolio of properties valued at more than two billion dollars, I've learned many methods that have brought success and longevity. Buildings must be designed and constructed thoughtfully with an unwavering commitment to safety. That is why a book such as this is of utmost importance to all developers and construction managers alike.

I've had the pleasure of knowing Matthew Bourque both personally and professionally for several years. I was introduced to him through Kelly McKeown, the daughter of the late James L. McKeown, who was an integral member of the Cummings Properties team and a close personal friend. Prior to Jamie's tragic, sudden passing in 1996, at age 41, he had a distinguished career at Cummings Properties, where he grew from leasing agent in 1979 to company president in 1990 and a founding trustee of Cummings Foundation.

I am still close with Jamie's wife, Denise McKeown, and her daughters, so I have gotten to know Matthew over the course of his relationship with Kelly. I had the honor of being by Matthew and Kelly's side as they got married. It has been a heartfelt pleasure to see their family grow with the additions of their son, Noah, and their daughter, Emma.

Our relationship soon grew to include a professional connection as well. Matthew's special expertise has been invaluable in helping us navigate fire prevention regulations, management of fire protection systems and overall building planning. Matthew brings a wealth of knowledge surrounding fire protection engineering while also bringing practical, effective solutions.

As we all know, fires can be devastating to occupants, buildings and communities, and fire safety is one of the most important aspects of construction. In our experience, fire prevention strategies go far beyond code compliance, with foresight and long-term safety in mind.

Fire safety must be considered at all stages – from initial design to construction to occupancy. If properly planned for and incorporated early, the fire prevention and life safety considerations for a project do not need to be a hindrance. As real estate developers, we certainly hold the responsibility to carefully monitor and influence the level of safety our buildings hold.

Matthew's fine new book, *Fire Prevention in Modern Construction: A Practical Guide*, is an essential work for engineers, property managers, construction managers, and all involved with construction, real estate, and public safety. This book offers a detailed examination of the complexities of fire prevention while offering best practices, case studies, and useful guides.

Matthew's studies of Fire Protection Engineering at Worcester Polytechnic Institute, his involvement with National Fire Protection Association technical committees, and his deep experience in some of New England's most modern and prominent buildings have all helped provide him with a well-rounded knowledge of fire prevention in buildings.

For all in the real estate and construction fields, this book is a must-read that may transform your approach and thinking to ultimately protect lives. I am honored to endorse this important work and look forward to witnessing the positive impact it will have on the industry for years to come.

Bill Cummings
Founder, Cummings Properties
Founder, Cummings Foundation
Author, *Starting Small and Making it Big* (2018)

Acknowledgments

I would like to acknowledge the many people who supported me while I wrote this book. It would not have been possible without the unwavering support of my family, friends, and colleagues.

First and foremost, I would like to express my gratitude to my family. Thank you to my wife, Kelly, for providing encouragement, patience and love throughout the whole process. Completion of this book would not have been possible without your help and support. Thank you to my son Noah and daughter Emma for being consistent and positive influences on me every single day. At the time of this book being published, Noah you are four years old and Emma you are two years old. I cannot wait for you to be old enough to read what I have written someday. You both make me so proud and will continue to do so as you grow. Everything I do, including writing this book, is for you both.

I would also like to honor my dog, Penny, who spent countless hours accompanying me while I wrote this book. Writing would not have been the same without her sleeping in her bed by my desk every single day. Though she passed before the completion of the book, she remained in my thoughts while I continued to write. I will never forget what a special and wonderful dog you were. Thank you for being the best friend I could ever have.

Thank you to my parents, Raymond and Susan Bourque, whose early lessons shaped who I became and continue to shape me today. The strong drive to work hard, do good and help others was instilled in me at an early age. Thank you to my siblings, Casey and Ryan, who have been there every step of the way throughout my life and career. You have both set such a positive example, both personally and professionally and have inspired me consistently. I am so lucky to have you both in my life.

I am extremely grateful to countless colleagues and friends who lent their expertise and support throughout the process. From providing reviews, feedback, ideas and overall support, I appreciate every single person, and your efforts have not gone unnoticed or unappreciated. Thank you for all your wise advice and encouragement.

I am very thankful for the support of Sadie St. Germain, of St. Germain Design Co., for the development of several diagrams in the book. I am so fortunate to have had your talent contribute to my book, and I appreciate your help so much.

Last, but not least, I would like to thank Taylor and Francis, Routledge and the incredible people assigned to this project. Thank you to my editing team, Emily Irvine and Sonia Tam, whose insightful feedback allowed me to refine the book and create a high-quality product. Thank you for believing in the project, the idea and in me.

Part I

Introduction to Fire Prevention in Buildings

Chapter 1

Background and History

1.1 Pre-21st Century Fire Safety Considerations

Building fire safety has made major advancements over the past century. Improvements have been mainly driven by updates to more stringent building and fire codes, resulting in improved building materials, installation of several fire protection systems, education surrounding fire safety and other factors. Even with these advancements, fires continue to tear through buildings.

New codes and regulations pose strict requirements regarding the construction materials permitted in buildings. They set limitations on the height and area of buildings. The codes require a number of life safety systems to be installed to set a minimum level of safety for occupants as well as property protection.

The earliest form of a building code can be traced back to The Code of Hammurabi from the Babylonian Dynasty in the 1700 B.C. timeframe. Rule #229 reads as follows: "If a builder builds a house for someone, and does not construct it properly, and the house which he built falls in and kills its owner, then that builder shall be put to death."[1] While this may sound extreme, it is a stark reminder of the responsibility that an owner and builder hold when constructing homes and buildings that people occupy.

Since those early days, building and fire codes have progressed as settlements became more condensed, and buildings grew in size. Large colonial cities such as Boston had building acts to set forth minimum building standards to prevent major losses from fire. Fires were much more frequent back then, with minimal life safety systems and without significant consideration given to construction materials.

New York's First Tenement Housing Act in 1867 was a result of the extremely poor conditions in tenement buildings. These acts require fire escapes and other minimum safety features. Over the 19th and 20th centuries, building and fire codes were developed and implemented in major cities across the United States and other large cities in the world. These codes were often the result of a catastrophic fire. Many large cities across the United States and other countries suffered from a "Great Fire." The Great Chicago Fire of 1871, and the Great Boston Fire of 1872,

DOI: 10.1201/9781032720210-2

and many other large-scale fires fueled the need for fire and building codes to be implemented.

Over the past century, we have also seen buildings go from having little to no fire protection system features to the present day, where buildings contain many engineered fire protection systems and other life safety considerations. Fire alarm notification systems and fire suppression systems were not common in the early 1900s. Passive fire protection systems, like fire walls and other fire or smoke-rated elements, were still new ideas that were in their infancy. The priority in fire prevention has shifted greatly in the way we design, construct and operate buildings.

The advancements in building and fire codes have set a minimum level of safety in buildings. However, this minimum level of safety is often based on the condition of the building at completion, when all designed fire and life safety systems are complete and active. When these buildings are under construction, they are extremely vulnerable.

During the 20th century, building and fire codes became more widely adopted and became a new norm when constructing buildings. The first edition of the National Fire Protection Association's (NFPA) 241, *Standard for Safeguarding Construction, Alteration, and Demolition Operations*, was released in 1968. This standard sets minimum fire safety thresholds for buildings under construction. This standard laid the groundwork for modern construction safety practices by providing requirements for fire prevention, hot work protocols and handling of combustible materials. This standard was developed in response to a growing threat of fires in buildings stemming from construction operations. Although this standard has been in place for over half a century, it was not widely adhered to nor enforced. Fast forward to the present day in the 21st century, and we are on the 14th edition of the code. There is also much more traction than there was in 1968.

Additionally, if modifications are made to the building throughout its lifespan without life safety in mind, this can jeopardize the building's performance in a fire. There has been a pattern showing the lack of consideration made by owners, developers, property managers and others who manage buildings day to day. They struggle to make informed, good decisions that do not diminish the building's level of life safety. The maintenance, repair and improvements carried out on buildings over their lifespan must be carried out with fire prevention and life safety in mind. Otherwise, the cumulative effects of negligence, cost-saving tactics and cutting corners could be a recipe for disaster.

There has been a major increase in awareness surrounding fire safety over the past several decades. Children are educated in fire safety in schools at an early age, participate in fire drills and often dedicate a week or month to fire safety awareness. Field trips to fire stations introduce children to the fire safety world. There have also been many organizations and resources that support education in this subject. To successfully achieve any goal, education is a very important starting point.

1.2 Current Day Fires

In this day and age, we still see devastating fires in a variety of places, even in the most advanced urban areas. NFPA's November 2024 Report on Fire Loss in the United States During 2023 summarized the following statistics regarding the impact fire has on the country[2]:

- Fire Departments responded to 1.39 million fires
- 3,670 civilian fire deaths
- 13,350 reported civilian injuries
- $23 billion in property damage

These staggering numbers were included in the report:

> *On average, a fire department responded to a fire somewhere in the US every 23 seconds in 2023. A home structure fire was reported every 95 seconds, while a home fire death occurred every three hours, and a home fire injury occurred every 52 minutes.*

Even in present-day times, fires continue to have a major impact on the country.

1.2.1 Current Day Construction Fires

Buildings under construction can be even more vulnerable than a completed building since life safety systems and other fire protection features are not yet in service or are impaired. Furthermore, construction processes can present fire risks.

According to the NFPA's October 2022 Report on Fires in Structures under Construction, there is an upward trend in the number of fires on construction sites over the past decade. The annual impact of construction fires is summarized with the following statistics[3]:

- 4,750 fires at buildings under construction
- 5 civilian fire deaths
- 62 reported civilian injuries
- $376 million in property damage

This book will explore several major construction site fires to understand their causes, devastation and most importantly, how they could have been prevented or lessened.

1.2.2 2019 – Notre-Dame Cathedral – Paris, France

In 2019, the well-known Notre-Dame Cathedral located in Paris, France, suffered a major fire when the building was undergoing substantial renovations. On April 15, 2019, at approximately 6:00 P.M., a fire broke out within the attic area of the building. The attic was mostly constructed of wood timber structural elements. The fire rapidly spread across the attic space, spread across the roof and quickly engulfed the iconic spire, collapsing it. Due to the historical significance of Notre-Dame, the impact of the fire on the city, country and world was enormous.

The official cause of the fire is still undetermined, although it is generally classified as accidental. There are several hypotheses of the cause of the fire, such as a short-circuited electrical wire, hot work, a cigarette and others.[4] The cathedral's fire alarm system detected the fire, although the system was not designed to automatically transmit the alarm signal to the fire brigade. It wasn't until a guard on-site investigated the location of the alarm and called the emergency services. The guard's first attempt at locating the fire was unsuccessful as they went to an incorrect location, which delayed the detection and the call to the fire brigade.

The renovations to the cathedral introduced several changes to the existing building: temporary equipment and materials, contractors and workers and modifications to the structure. Work activity occurring in the centuries-old wood-timber attic space presented a lot of risk. The wood elements became very dry over time, increasing their combustibility characteristics.

The rebuilding of Notre Dame took over 5 years to complete. It was estimated that 2,000 stonemasons, carpenters, sculptors, roofers, other skilled labor and engineers worked on the project that cost about $800 million. Nearly one billion dollars was raised by donors around the world, highlighting the impact this fire had on many people (Figure 1.1).

Figure 1.1 2019 Notre Dame Fire. Photograph by Troude, Godefroy. *Notre-Dame en feu, 20h06*. April 15, 2019. Wikimedia Commons. Licensed under CC BY-SA 4.0.

1.3 A Solution to the Problem

Many construction fire safety procedures and protocols found on construction sites were implemented as best practices before the formal development of a specific plan to address these items was created. The development of a comprehensive construction fire safety plan has become customary in many places during the construction phase of a project. In some jurisdictions, these construction fire safety plans are mandatory and are even a condition of the issuance of a building permit. These plans address the hazards involved with the construction or renovation of a building and propose mitigation strategies to eliminate these risks. In jurisdictions that adopt NFPA 241, the creation of a construction fire safety plan is a code requirement, although the enforcement of the plan may vary. However, there is a growing trend sweeping across the United States and other areas, bringing light to the NFPA 241 standard and the need for a comprehensive construction fire safety program. When cities and towns suffer terrible construction fires, destroying buildings and sometimes loss of life, there is a common consensus that there needs to be more done to prevent fires. The answer is typically stricter enforcement of construction fire safety practices. The individual responsible for implementing the construction fire safety plan is the Fire Protection Program Manager, often referred to as the FPPM.

References

1. Hammurabi, *The Code of Hammurabi*, trans. L. W. King (Chicago: The University of Chicago Press, 1903), §229.
2. Hall, Shelby. *National Fire Protection Association (NFPA) Report on Fire Loss in the United States*. November 2024. https://www.nfpa.org/education-and-research/research/nfpa-research/fire-statistical-reports/fire-loss-in-the-united-states.
3. Campbell, Richard. *Fires in Structures under Construction*. Quincy, MA: National Fire Protection Association, October 2023. https://www.nfpa.org/education-and-research/research/nfpa-research/fire-statistical-reports/fires-in-structures-under-construction.
4. Le Monde. "Five Years after Notre-Dame Fire, No Culprit Has Been Found as Cathedral's Long-Awaited Reopening Nears." *Le Monde*, November 9, 2024. https://www.lemonde.fr/en/france/article/2024/11/09/five-years-after-notre-dame-fire-no-culprit-has-been-found-as-cathedral-s-long-awaited-reopening-nears_6732266_7.html.

Chapter 2

Introduction

2.1 Introduction

Considerations for fire prevention and life safety must be prioritized at each stage of a building's life cycle. In early planning and design, fire protection systems must be thoughtfully integrated into the building's design to prioritize future life safety of occupants and provide property protection. Additionally, systems must be designed without being an operational or aesthetic hindrance. During construction, fire prevention must be closely managed to overcome the significant fire hazard risks associated with this phase. When construction is complete and the building becomes operational, there are many new life safety considerations that must be kept at the forefront. Therefore, building fire prevention is continuous throughout the entire lifespan of a building; rather than only being considered during the construction phase.

2.2 Fire Protection Basics

Four components must be present for a fire to occur: fuel, oxidizer, heat and an uninhibited chemical chain reaction.[1] If one of these items is not present, there cannot be a fire. This is commonly referred to as the fire tetrahedron. The fundamentals of building fire safety, similar to fire prevention in general, rely on limiting or removing one or more of these four fire contributors. Fire prevention on a construction site and operating buildings primarily focuses on the two main controllable components: fuel and heat (Figure 2.1).

The construction fire safety plan will identify potential fuel sources in the building, potential heat sources from construction processes and provide a mitigation plan to minimize or eliminate these hazards. There are many fire hazards found in buildings, both for the regular operation of a building and those supporting construction operations. Electrical systems, appliances, heating systems, combustible contents, flammable and combustible liquids and smoking materials could all be hazards if not properly managed. Construction activities that produce heat or sparks, temporary heaters, and other equipment that produce heat could be hazardous if not carefully managed. The forthcoming chapters will explore these elements in greater detail.

DOI: 10.1201/9781032720210-3

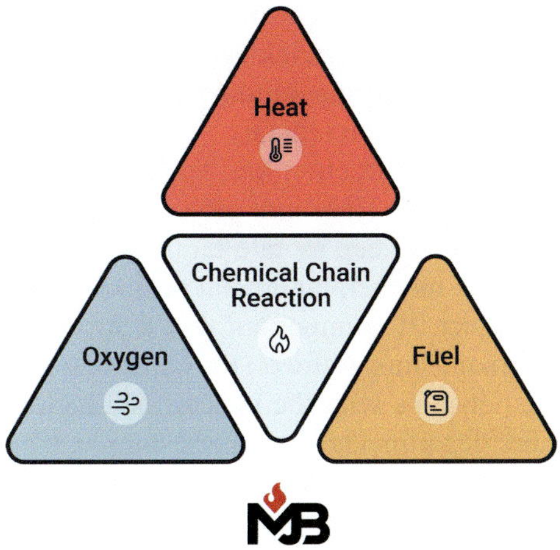

Figure 2.1 Fire Tetrahedron. Image illustration by St. Germain Design Co.

2.3 Importance of Fire Prevention

Construction fire safety on the jobsite is critical for many reasons. First and foremost, it is important to protect the lives of occupants and workers who are within the building. When a building is under construction, there are many people who spend a significant amount of time at the building. This includes owner and developer personnel, general contractor's staff, subcontractors and their workers, inspectors, architects, engineers, prospective tenants, property managers and other parties. The familiarity of the building also varies amongst this group. Someone who works in the building 40 hours each week will have a different level of knowledge of the layout, exits and fire protection systems than someone who performs a quick visit once a month. The minimum level of safety established by the construction fire safety plan must account for these differing levels of familiarity.

Often, construction projects take place in existing, occupied buildings. In these instances, it is extremely important to ensure that the construction activities occurring do not pose a fire risk or diminish the building's life safety level in any way. If the building's fire protection systems are jeopardized due to the construction activity elsewhere in the building, the building and its occupants could be in severe danger should a fire event occur.

Another reason construction fire safety is essential is for property protection. Buildings that suffer fires endure huge property losses. There are direct impacts of the fire where construction materials, structure, finishes, and equipment are damaged. These areas usually require significant rework and replacement of most materials. Almost equally damaging are the indirect damages of the fire from the byproducts of combustion. Areas outside the fire zone often will sustain damage from smoke and other toxic gases from the fire. The smoke and other combustion byproducts will be absorbed in many materials found on a construction site. Often, there is also significant water damage stemming from fire hoses used by firefighters to control the fire, or from fire sprinklers. Specialty testing companies often come to the site to test all areas of the building to see what items must be treated, removed and replaced, or which can remain.

When a project suffers a construction fire, there is also the loss of morale factor and a strain on relationships. Tight-knit project teams are formed on construction projects. It is important to keep morale high on the project; to keep workers motivated to work in a safe manner, hit crucial project milestones, and to provide a high-quality work product. It is important that a strong relationship is formed across internal and external teams. A fire on the construction site can cause tension in these important relationships with the general contractor, owner, design team and the building and fire departments alike. For example, a building or fire inspector may become less cooperative or less accommodating and take a more conservative approach for the remainder of the project, or even on future projects with these parties. Often, fingers are pointed to prevent blame on certain groups.

A fire can cause substantial delays to a project. These delays can stem from rework, stop work orders, and material supply chain issues of replacement materials or equipment. The procurement of construction materials and equipment is planned early in the project to ensure items are on-site when necessary to keep the project moving in an efficient style. When an unexpected event like a fire occurs, critical building items like custom curtainwall systems, chillers, generators, switchgear and other long-lead time items could be damaged. These items need to be refabricated and sent out to the site, often holding up completion of major building systems. Many areas of the building will need to be stripped down to the structure and replaced with new materials. This often pulls workers from other areas of the building to reprioritize them to replace what was damaged. Another factor is stop-work orders. These directives can come from authorities having jurisdiction (AHJ), including fire or building officials, insurers or other stakeholders. Fire departments may pull permits until the project reevaluates their construction fire safety plan and provides a correction plan based on the events that took place. Insurers may require a site visit and investigation to determine the cause and a plan on how this will be prevented when construction resumes. Investigations, reporting and new permits are time-consuming processes. There could be days, weeks, months or sometimes years before the project resumes. These significant delays prolong the construction project and push out the occupancy, which results in a delay in bringing in revenue. This causes major financial strains on the investors and owners.

The negative impacts of a construction fire are multidimensional and create an enormous challenge to overcome to complete the project. The project team must pivot from the previous trajectory and deal with several complications in the aftermath of a fire. Construction fire safety must be the top priority on a construction project. A safe project will be a successful project. The chance of a construction fire occurring can be significantly reduced if a comprehensive construction fire safety plan is developed and implemented for the project. The Fire Protection Program Manager (FPPM) plays an important role in ensuring the contents of the plan are being carried out. The

FPPM can influence the development of the plan and can make recommendations on construction materials or processes that reduce the fire hazard risks.

There is no greater protection from fire than preventing it in the first place. As Benjamin Franklin said, "an ounce of prevention is worth a pound of cure."[2] This sentiment could not be truer in the world of fire prevention.

2.4 My Road to Fire Protection Engineering

When most people think of engineering, they typically think of mechanical, civil, electrical, chemical or software engineering. There is no doubt that the fire protection engineering world is a niche field. With only a select few institutions offering education in the field, the career path often flies under the radar. Some are led to it by a relative in fire protection engineering or a family in the fire service industry. However, most, like me, are led to it by chance.

I grew up in Nashua, New Hampshire, a mid-size city about an hour north of Boston, Massachusetts. My interests always revolved around construction, math and science. At a young age, building with blocks and Legos was my most favored activity. As I got older, construction drew my interest. My dad is a lifelong carpenter, specializing in finish carpentry for over half a century. I would often find myself working on school vacations with him. Learning the ins and outs of woodworking from an expert. I loved everything about construction sites: the sights, the smells of fresh cut wood, drywall and paint, and the satisfaction of seeing tangible results at the end of a hard day's work.

When I was about 13 years old, I started a lawn mowing venture with my friend. We took on more than 25 lawns around my neighborhood and city. This continued until I was about 18. While it was hard work, it also taught me many valuable lessons in entrepreneurship, customer relations and business that continue to shape me today. It taught me the importance of being reliable. It also gave me the opportunity to make my own money at an age where it might otherwise not have been possible.

While I did enjoy the hands-on work I experienced with working construction and landscaping, I knew this could be a challenging path long term. Engineering seemed to be the most logical alternative. Engineering, as I understood it at a young age, involved designing and building things or structures and that seemed fascinating to me. So, starting at age 13, I decided I was going to be an engineer.

My brother, Ryan, who is five years older, also had similar interests. He studied civil engineering at Worcester Polytechnic Institute (WPI) in Worcester, Massachusetts. So, when the decision came on my plans after high school, the choice was clear – I would go on and study engineering. I applied to a handful of local schools in Massachusetts offering engineering programs. When the letter of acceptance came from WPI, I was ecstatic and made the easy decision to attend.

I entered WPI and enrolled in the mechanical engineering program, which was a broad engineering discipline where you could later veer off into many areas of focus. The first couple of years involved a fundamental math and science curriculum with many chemistry, physics, calculus classes, and other math disciplines. Thermodynamics, heat transfer, and fluid mechanics were next up. During my first couple of years in the program and understanding the outlook for the rest of the mechanical engineering path, I second-guessed myself and thought that I might be in the wrong place. My interests were back in construction, buildings and those tangible and relatable topics. I even consulted an advisor to see if I should switch to civil engineering. I soon learned of a different program at WPI, Fire Protection Engineering. This program involved the detailed study of the fundamentals of fire, fire protection systems, and the prevention of fires using engineering principles. Much of the focus was on fire prevention in buildings, which tied back to construction. The program was a graduate program where many undergraduate degrees nicely meshed with it. I quickly applied to the program and was admitted. I also studied both the mechanical engineering and fire protection engineering curriculum requirements to see how I could complete both degrees as efficiently as possible. The graduate-level fire protection engineering courses double-counted for my mechanical engineering advanced courses. When I graduated with my undergraduate degree in four years, I only had two more courses required to achieve the Master of Science Degree in Fire Protection Engineering.

Upon graduation with my undergraduate degree, I sought full-time employment. The four years at college racked up a significant amount of student loans and I was eager to start making money and kick off my career. I took a job at an engineering company based in Boston. The company served as the traditional MEP/FP-type design firm, plus they had a fire protection engineering group, referred to as the "code group," where I landed. I was able to finish my fire protection engineering graduate program over the next year. One year after graduating with my undergraduate degree, I graduated again with my master's degree in Fire Protection Engineering.

Over the next several years, I gained a significant amount of experience working with licensed fire protection engineers on a wide range of projects. I was involved in existing building evaluations, often producing code reports summarizing the applicability of code requirements to projects. I gained extensive experience inspecting buildings and developing recommendations to owners, developers, architects and other stakeholders. I was involved with the design and fire modeling of smoke control systems and the testing of these systems. I was also exposed to fire sprinkler design. I supported construction projects and helped develop construction fire safety plans.

The path of becoming a licensed professional engineer is a rigorous journey. It often includes obtaining minimum education requirements, work experience and passing two difficult exams. The education piece typically involves completion of an accredited engineering degree, often a bachelor's degree in engineering and/or a master's degree in engineering. Full-time work experience must be completed working directly with a licensed professional engineer. The first major exam is the Fundamentals of Engineering Exam, often referred to as the "FE" exam. When I took this exam in 2013, it was an in-person, pencil-and-paper exam that spanned over the course of eight exhausting hours.

When education, work experience and completion of the FE exam are met, the next and final hurdle is the Principles and Practice of Engineering Exam, often referred to as the "PE" exam. Work samples, references, and transcripts accompany the robust professional engineering exam application. When approval by the board is given, applicants can sign up for the exam that is offered just once a year for fire protection engineering. This exam was also a full-day, in-person, pencil and paper exam. Exam questions covered all areas of fire protection engineering. Unlike some exams where there is an established pass rate, this exam was evaluated each time it is administered based on results from each question, patterns and evaluated amongst other peers taking the test. I was fortunate to pass both the FE and PE exams on the first attempt. In December of 2016, I became a licensed Fire Protection Engineer.

At the same time that I was obtaining my professional engineering license, I was also approached with a new opportunity. A significant development was underway in the Seaport District in Boston. This development spanned many city blocks, covering more than 20 contiguous acres. New mixed-use buildings, parks, parking garages and roadway improvements were all included in this master redevelopment. These projects were big, complex and were in the spotlight since this new neighborhood in Boston was seeing significant growth. The buildings involved multiple retail tenants within the lower levels, in addition to subgrade parking garages. The space in the tower above were often residential, office or life science uses. The city's building and fire departments needed reassurance that this development would be managed carefully to successfully coordinate many different moving pieces. That's where my role was created. I started in January 2017 as the FPPM for the Seaport Development.

2.5 Targeted Audience

The purpose of this book is to provide a comprehensive guide to safeguarding structures from fire hazards at all stages of a building's life cycle, from early design, through construction, to an operating building. It is intended to be a resource for those serving in the role of the FPPM. The information, strategies and prevention approaches are not only useful for those in fire prevention-focused roles, but rather for all people who are involved with construction or the management of buildings. This includes construction managers, superintendents, property managers, building engineers, tradesmen and tradeswomen. Fire prevention must be considered from all angles, as many decisions influence a building's level of safety.

References

1. Firesafe.org.uk. "Information about the Fire Triangle/Tetrahedron and Combustion." Last modified February 10, 2022. https://www.firesafe.org.uk/information-about-the-fire-triangletetrahedron-and-combustion/.
2. Franklin, Benjamin. *On Protection of Towns from Fire* (Philadelphia, 1735). https://founders.archives.gov/documents/Franklin/01-02-02-0002.

Chapter 3

Applicable Codes and Standards

3.1 Codes and Standards Introduction

There are many codes and standards that influence the construction of a building. Building and fire codes pose strict guidelines on allowed construction materials, finishes, exiting requirements, roofing materials, fire ratings and life safety systems. It is important to understand the purpose of codes and standards. The requirements found in these documents set the minimum design and construction requirements to ensure a safe and structurally sound building which therefore protects both life and property alike.

Many decisions about how a building is designed and constructed are governed by codes and standards. When an option is presented for a component of the building, it is often asked, "is this code compliant?" Going down a path with a design that is not code compliant is obviously not very effective. Or, sometimes, the approach of "what is the code minimum?" is used. By doing the absolute code minimum, you can achieve code compliance without going above and beyond. This common approach is the most efficient way when considering financial and code compliance goals. It may be suitable for some parts of a building's design and not suitable in other areas where increased fire prevention, energy efficiency, or building performance is desired.

Most areas of the United States utilize model codes by the International Code Council (ICC). The ICC creates model codes such as the International Building Code (IBC) and the International Fire Code (IFC). These codes are continually updated on a regular basis. These model codes are then adopted by states, cities and towns-often with their own specific amendments, but the basis of the code is derived from the ICC model codes. Canada, along with several countries in Europe, often use codes, acts and building standards developed by governmental agencies. These codes are adopted by provinces, states, and local governments.

The National Fire Protection Association (NFPA) produces over 300 codes and standards. Many of these codes and standards are similarly adopted, sometimes with amendments, by many states, cities and towns. They are also referenced and scoped into the ICC model codes. Therefore, when the IBC is in effect in a certain jurisdiction, many NFPA codes and standards must be met as well.

DOI: 10.1201/9781032720210-4

Some municipalities have their own specific fire prevention code or regulations that are not derived from a model code. These are often seen in larger cities and usually evolved from decades past. The requirements found in these codes are sometimes more stringent than the requirements found in the building or fire code adopted. As model codes have predominantly taken over, these specific fire prevention codes or regulations have slowly disappeared or become obsolete (where requirements are found in other adopted codes).

Codes and standards vary from state-to-state, city-to-city and town-to-town. Therefore, it is important to research all applicable codes and standards that will apply to the project. These documents are updated periodically. For this reason, it is also important to ensure you are utilizing the correct version of the code. Even if there is a newer code or standard issued, the project where the building is being constructed may adopt a previous version. Adoption of new codes and standards often happens at the state level. This includes many reviews, public hearings, updates and other promulgation steps. Therefore, it is not uncommon for this process to take a year or two. It becomes the foundation of the design and safeguard the building while under construction. It is extremely important that the basis of the codes is established and accurate.

On complicated or challenging projects, it is beneficial to consult with a building and fire code specialist who can provide guidance on the applicability of various codes and standards specific to the project. On more straightforward projects, it may be possible to navigate them without the support of a code consultant. Most jurisdictions will have a portion of their website dedicated to building and construction-related information. The website will often summarize building and fire codes that are currently in effect and will include amendments, if any, to these referenced codes (Figure 3.1).

3.2 Prescriptive and Performance-Based Requirements

Requirements found in codes and standards can be prescriptive or performance based. Prescriptive requirements include specific parameters, including materials, sizes, dimensions and quantity. Whereas performance-based requirements set goals and desired results. The example below shows differences in prescriptive and performance-based requirements:

Prescriptive: a 1-hour fire rated assembly must be constructed with 5/8″ Type X drywall on both sides of a 2′×4″ framed wall. Fastening, stud spacing and other parameters are outlined.

Performance-based: 1-hour fire rated assembly must be provided in accordance with a nationally recognized standard. This could be achieved in many ways. It could be achieved using the arrangement in the prescriptive method. A CMU (concrete masonry unit) wall can be utilized. A fire-rated glass, or sprinkler protected glass could meet the performance criteria as well. Other equal arrangements could be used as long as they meet the goal criteria of a 1-hour fire rating.

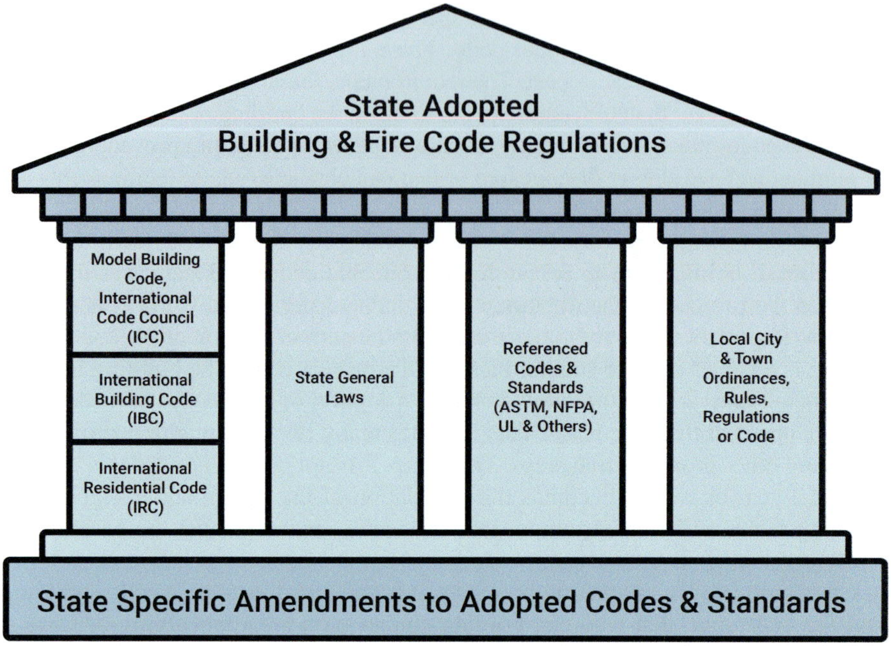

Figure 3.1 State Adopted Building & Fire Code Regulations. Image illustration by St. Germain Design Co.

The performance-based requirements provide the designer with significantly more flexibility in achieving the goal established. This becomes important when there are other goals of the project that are top priority like aesthetics, material availability and constructability. However, with the advantage of flexibility comes the responsibility to develop an acceptable solution to the set goal.

3.3 Application of Code Requirements

Many code requirements are specific to occupancy classifications and construction types. Further requirements or exceptions are provided based on the fire protection systems present. Therefore, it is important to understand code classifications for your project. This ensures you are accurately applying requirements found in the code.

3.3.1 Occupancy Classifications

Occupancy classifications of a space or building are one of the main categories that must be established. The occupancy classifications are based on the expected use of the space or building. The main categories of occupancy classifications are tied to classifications found in the IBC,[1] and NFPA codes, NFPA 101 – Life Safety Code,[2] NFPA 5000 – Building Construction and Safety Code.[3]

Assembly occupancies (Use Group A) – these types of uses involve the gathering of occupants in a building for various purposes such as dining, entertainment, religious purposes, sporting events or other congregations of people. The IBC and NFPA codes have several different distinctions within the assembly classification to further classify it. These occupancies typically have high occupant loads due to the nature of the spaces.

Business occupancies (Use Group B) – these spaces are used for professional services such as offices and work areas in addition to transactional services such as banks, dry cleaners, post offices and the like. Higher education occupancies are also typically considered business occupancies.

Educational occupancies (Use Group E) – these include buildings used for educational purposes, typically up until grade 12. Higher education occupancies would typically fall under business. Early education and daycare facilities are sometimes a sub-category of this use group.

Factory occupancies (Use Group F) – these spaces include the manufacturing, assembling of parts, equipment and other elements and are often sub-categorized based on materials and processes.

High-Hazard occupancies (Use Group H) – these spaces involve the storage or use of hazardous materials such as flammable and combustible liquids and gases, combustible dusts and other materials that could present a risk to occupants, workers and/or the property. This group is sub-categorized into different groups based on the hazardous materials.

Institutional occupancies (Use Group I) – this classification is used in cases where occupants may not be capable of self-preservation and require assistance from workers, aids, and employees to safely evacuate. Occupancies could include daycares for young children and adults, detention centers, detox centers, hospitals, nursing homes and other similar uses.

Mercantile occupancies (Use Group M) – these uses include retail-type occupancies that sell and showcase products and other goods. It can include department stores, grocery stores, showrooms and other similar stores or shops.

Residential occupancies (Use Group R) – these uses include buildings that are used for living and sleeping purposes. This category is further split based on the nature of the building. It includes transient-type uses such as hotels, motels and other uses. Multi-family buildings such as apartments, condominiums, townhouses. It includes one- and two-family homes. It also includes assisted living, boarding homes and other unique arrangements.

Storage occupancies (Use Group S) – this use includes storage of materials, equipment, packaging, vehicles, products and other items. It is further sub-categorized based on the type of items in storage.

Utility and Miscellaneous occupancies (Use Group U) – this classification is for the uses that don't cleanly fit into one of the other use group classifications. This may include agricultural buildings, sheds and barns and uniquely used structures.

3.3.2 Construction Types

The construction type classification of a building is dependent on the material and fire ratings of the main structural components of the building, such as framing, bearing and non-bearing walls, exterior walls, roof, and floor assemblies. The following table summarizes the most common construction types categorized by codes.

Construction Type	*Description*
Type I	Type I Construction is further split between Type IA and IB subsets. This type of construction consists of noncombustible materials, with limited combustible materials for specific applications. Type IA construction includes a 3-hour fire rating on many structural components, such as columns and beams and a 2-hour rating on floor assemblies and roofs. Type IB construction includes a 2-hour fire rating on many structural components, such as columns and beams.
Type II	Type II Construction is further split between Type IIA and IIB subsets. This type of construction consists of noncombustible materials, with limited combustible materials for specific applications. Type IIA construction includes a 1-hour fire rating on many structural components, such as columns and beams, floor assemblies and roofs. Type IIB construction includes unprotected structural components (no fireproofing, no enclosure).
Type III	Type III Construction is further split between Type IIIA and IIIB subsets. This type of construction consists of a combination of noncombustible, fire-rated exterior walls, such as masonry with combustible interior framing, such as wood-framed construction. Type IIIA includes a 2-hour fire-rated exterior wall and a 1-hour fire rating on many interior structural components, such as primary structural frame and bearing walls and floor and roof construction. Type IIIB includes a 2-hour fire-rated exterior wall with unprotected interior combustible framing, such as wood-framing.
Type IV	Type IV Construction includes heavy timber construction. Exterior bearing walls are 2-hour fire-rated and the interior primary structure meets specific dimensional requirements of wood or other composite materials.
Type V	Type V Construction is further split between Type VA and VB subsets. This type of construction consists of a combustible interior structure, such as wood-framing. Type VA includes a 1-hour fire rating on most elements such as exterior bearing walls, primary structure, floor and roof. Type VB includes a combustible, wood-framed structure without any protection. Type VB Construction is the least restrictive construction type. Most one and two-family homes are constructed as Type VB construction.

3.4 Fire Protection Systems Present in Building

It is important to understand what fire protection and life safety systems are installed in the building, as several code requirements are based on these parameters. Subsequent chapters will go into further detail on these systems. However, it is critical to understand if the following systems are provided in the building and to what extent:

1 Fire Alarm System (Chapter 6)
   a Full coverage or partial system
   b Determine if addressable or conventional system
   c Determine capabilities of system (voice evacuation)
   d Determine type of detection and notification appliances
2 Fire Sprinkler System (Chapter 7)
   a Fully sprinklered or partial system
   b Determine standard system was designed to
   c Determine design classification of system
3 Standpipe System (Chapter 7)
   a Determine if standpipe system is provided
   b Determine what type of system is provided
   c Determine floors of building served by system
4 Smoke Control Systems (Chapter 8)
   a Determine if smoke exhaust systems, pressurization systems, smoke lobbies or other smoke control systems are present

3.5 Applicable Codes and Standards During Construction

Building and fire codes significantly influence the building's design to create a safe building when construction is complete. However, until construction is complete, the building traverses through a very vulnerable state. During construction, fire protection systems are yet to be livened up, structures yet to be protected with gypsum or a spray-on fireproofing product, and rated stairwells yet to be enclosed in rated construction. Additionally, wood and other combustible materials are used for temporary conditions. Therefore, there needs to be a focus on the hazards presented during the construction process.

NFPA 241 – *Standard for Safeguarding Construction, Alteration, and Demolition Operations* is a standard developed by the National Fire Protection Association. This standard specifically addresses those hazards presented during construction, including alteration and demolition processes. This standard, like other codes and standards, provides both prescriptive requirements and performance-based requirements.

3.6 Permitting

In addition to codes and standards, jurisdictions will also implement processes, procedures and permits for various construction-related activities. Most towns and cities require the following basic permits:

- Building Permit
- Electrical Permit
- Plumbing Permit
- Mechanical/HVAC Permit

The following permits are specific to fire prevention:

- Fire Alarm Permit
- Fire Sprinkler Permit
- Smoke Detector Bagging Permit
- Construction Fire Safety General Permit
- Hot Work Permit
- Dumpster Permit
- Fuel Storage Permit

Some or all permits (or others not included here) may be applicable based on the location of the project. It is important to familiarize yourself with local permitting requirements. Each jurisdiction has slightly different processes and permitting systems. It is important that these permitting processes are followed correctly to ensure the project is abiding by all applicable regulations. Inspections will be performed at various project milestones that are determined by the AHJ. The AHJ is responsible for enforcing the applicable codes and standards by way of plan reviews and inspections. It is crucial to build a relationship with the various AHJs on the project to understand their expectations. By establishing a constant line of communication, you can keep them updated on construction status, run questions by them, and build a level of trust that will be especially important when unique situations arise.

References

1. International Code Council. *International Building Code*. 2021 ed. Country Club Hills, IL: International Code Council, 2021.
2. National Fire Protection Association (NFPA). *NFPA 101: Life Safety Code*. 2021 ed. Quincy, MA: National Fire Protection Association, 2021.
3. National Fire Protection Association (NFPA). *NFPA 5000: Building Construction and Safety Code*. 2021 ed. Quincy, MA: National Fire Protection Association, 2021.

Chapter 4

Construction Fire Safety Plan

4.1 2023 – 239 Unit Residential Building, "SouthPark" – Charlotte, North Carolina, U.S.A.

On May 18, 2023, a fire broke out at an apartment complex in Charlotte, North Carolina, taking the lives of two construction workers. This large residential complex was located in the SouthPark neighborhood of Charlotte. The building was constructed as a podium-style building, with non-combustible construction on the lower levels, while the upper levels were constructed with combustible wood framing. A common construction and design approach was utilized to maximize the allowable height and area of this building. We will take a detailed look at how these events played out on the day of May 18th, 2023, to understand what exactly unfolded by referencing a lawsuit filed in the North Carolina state court[1]:

4.1.1 Timeline

7:00 a.m. Demonte Sherrill and Reuben Holmes arrive at the project to begin work installing windows.

7:30 a.m. A worker starts a generator located on a mobile trailer alongside spray foam insulation equipment and materials. This trailer was located on the parking deck at grade, directly below the five floors of wood construction above. Worker starts spray foam installation on floor level(s) above.

8:00 a.m. The worker returns down to the parking level to check on the mobile trailer containing generator and spray foam equipment after the spray gun stops operating. The worker observes flames coming from the back of the trailer. The worker unsuccessfully attempts to use a fire extinguisher to put out the fire. The worker calls the supervisor but does not call 911. The worker left the area to find a supervisor to notify him of the fire.

8:55 a.m. The Mill Creek employees are becoming aware of the fire. The worker attempts to use fire extinguisher in trailer to extinguish the fire, but fire continues to grow. The worker reports the fire to supervisor.

9:02 a.m. 911 call is placed by the developer, Mill Creek Residential.

9:07 a.m. Charlotte Fire Department first apparatus arrives onsite.

DOI: 10.1201/9781032720210-5

9:09 a.m. Mill Creek superintendent tells Captain B.W. Benson of the Charlotte Fire Department that the building has an active standpipe that could assist the fire department by providing a water source from within the building.

9:11 a.m. Firefighters attempt to connect to what they thought was a standpipe.

9:14 a.m. Firefighters learn workers are trapped on the fourth and sixth floors. The fourth floor has good visibility. They climb to the sixth floor where screams of "help" are heard.

9:16 a.m. Firefighters inside the building realize no standpipe exists.

9:18 a.m. Demonte Sherrill, trapped on the sixth floor with Reuben Holmes, begins to plead for help on Facebook Live.

9:19 a.m. The construction crane operator, who ultimately helped rescue more than a dozen workers, unsuccessfully tries to save Sherrill and Holmes by lowering the crane basket.

9:20 a.m. Firefighters enter the construction site's only stairway to attempt to rescue Sherrill and Holmes.

9:22 a.m. Firefighters reach the sixth floor, where smoke hangs at eye-level. The crew begins crawling. Just as an order that they evacuate is issued, they hear a cry for help. They reach a hallway where a captain bangs the floor and yells for the workers. The trapped men bang on floors and walls in response. The captain says the workers sound close, but firefighters could not find them. The crew moves to evacuate but gets lost due to low visibility. They call a mayday but eventually find the one set of stairs on their own.

9:26 a.m. Sherrill ends his Facebook Live broadcast. Keith Suggs, owner of Prestige Windows and Doors, which employed Sherrill and Holmes, calls and speaks with Sherrill. He tells Suggs that they are trapped.

9:37 a.m. Firefighters who tried to save Sherrill and Holmes exit the burning building, as ordered. The five-alarm blaze continues to spread.

4.1.2 Aftermath and Investigation

The fire prompted a huge response from the Charlotte Fire Department, which included around 90 personnel and reached 5 alarms. Fifteen workers had to be rescued from the building after becoming trapped. News and media coverage were also very significant; news agencies in Charlotte and beyond covered this developing story (Figure 4.1).

A diesel-powered generator caught fire and quickly ignited the spray foam insulation that was adjacent to the generator. The location of the equipment within the building footprint led to a fire that could quickly develop with no fire suppression to combat it.

Two construction workers, Demonte Sherrill and Reuben Holmes, were trapped on the sixth floor of the building. They made several attempts to notify first responders and their employer of their whereabouts by screams of help, Facebook live stream and phone calls. However, these attempts would not be enough, as they succumbed to the fire and smoke before rescue efforts could reach them. It was

Figure 4.1 SouthPark Construction Site Fire on May 18, 2023. Photograph by Charlotte Fire Department.

reported by the Charlotte Fire Department that the fire's temperature exceeded two thousand degrees Fahrenheit.

In the following months, the families of the victims filed lawsuits against the developer and other potentially responsible parties for failures and negligence that led to the deaths of the two workers.

OSHA, the Occupational Safety and Health Administration, also conducted its investigation of the incident. This is customary anytime there are construction accidents or fires that result in any casualties. It was discovered that the upper levels of the building were only provided with a single exit by a stairwell that extended up one end of the building. Only a week or two prior to the fire, a secondary temporary exit stair was deconstructed, which was in the vicinity of the two trapped workers. The only exit stair was over 400 feet away from Sherrill and Holmes, with limited exit signage.

A Mill Creek superintendent allegedly proposed that the location of the mobile trailer containing a generator and spray foam insulation equipment be within the parking level.

The building did not have a functional fire protection standpipe. There was misleading information communicated to the fire department that there may have been; however, they discovered after several minutes that there was not one.

There was also no established communication system to notify workers of a fire or other incident. The notification to workers in the building was limited to verbal communication and shouts of "fire!"

A construction fire safety plan (CFSP) and emergency response plan and procedures were not developed and implemented. The culmination of fire prevention malpractice on the construction site led to a deadly incident that should have been avoided or minimized to prevent the unnecessary loss of life.

4.1.3 Charlotte Fire Reflection

As will be explored in the forthcoming chapters in this book, construction fire safety must be prioritized and a constant focus throughout the entire construction process. It is critical that internal combustion-powered equipment is located in a pre-determined, designated, safe area. A means of notifying workers and occupants of an emergency by way of a fire alarm system, even if temporary, must be provided. Rapid, established communication channels when a fire is detected must be carried out. It is paramount that workers and supervisors are aware and educated on the fire protection systems on-site to correctly communicate their status to first responders, such as standpipe systems. All areas of the building must be provided with sufficient means of egress. All of these items must be carefully thought out, developed into a detailed plan, and implemented by a qualified person or team.

4.2 Safety by Design

Without intentional, consistent efforts, fire prevention within a building or construction site is not possible. Fire prevention and safety practices must be integrated into the work. It cannot be occasionally considered as a reactionary measure to appease authorities having jurisdiction (AHJs). Many can fall into survivorship bias. While people may not have experienced a fire or life safety issue because of their actions, it doesn't mean their continued actions won't. You can get lucky sometimes, but you cannot get lucky all the time.

To sustain fire prevention long-term, strategies must be established early, reviewed and evaluated often, and adjusted accordingly. This evolving process will ensure that the proposed fire prevention methods continue to be a suitable solution to the ever-changing construction site and building.

4.3 Construction Fire Safety Plan

A CFSP is a comprehensive, project-specific plan that identifies potential fire hazards associated with the planned construction activities and provides mitigation strategies to reduce the risk of fire on the construction site. The plan outlines specific procedures to be followed when impairing life safety systems, conducting hot work, and other fire prevention best practices. These plans are also sometimes referred to as "NFPA 241 Plans" or "241 Plans," as they refer to the main influencing standard, NFPA 241, *Standard for Safeguarding Construction, Alteration, and Demolition Operations*. They are sometimes referred to as fire prevention plans. All these names are generally synonymous with the CFSP. These plans are developed by qualified individuals who are knowledgeable in building and fire codes, construction processes, fire protection systems, hot work, impairment procedures and general fire safety.

This chapter gives a high-level summary of the contents of a CFSP and how they are developed. Subsequent chapters will dive into the various main sections of the plan. Information on what content must be included, how the FPPM manages it, and common challenges and best practices will be reviewed.

A common question that is asked is "when is a construction fire safety plan required?" In jurisdictions that adopt NFPA 241, typically from a building or fire code reference, a CFSP is required to be developed for all construction projects, regardless of size. Many are surprised to learn this, as many believed this was only required on large-scale projects or when requested by building or fire departments.

Single and two-family homes that are scoped into residential building codes may not be subject to NFPA 241. However, these structures are almost exclusively constructed out of combustible materials and consideration should be given to prevent the risk of fire during construction.

The CFSPs are as complex as the project is complex. On small-scale, simple projects, these CFSPs can be developed by the general contractor, owner, or other project stakeholder. These plans may also be straightforward and minimal. For larger and more complicated buildings and fire protection systems, these plans are usually developed by a fire protection engineer or specialist within the construction fire safety industry.

Regardless of the complexity of the project, the approach for developing the CFSP remains the same:

1 Identification of fire hazards
2 Mitigation strategies to eliminate or reduce hazards
3 Management of fire protection systems
4 Defined roles and responsibilities of key stakeholders and communication

4.4 Identification of Fire Hazards

Common with any risk assessment, fire hazards must be examined from both a severity and probability perspective. Best-case scenarios, expected outcomes and worst-case scenarios should all be considered. Past experiences should also be used as lessons learned to prevent repeat issues. The duration of construction activities within the building influences the probability of an event from happening.

There can be several fire hazards present on construction sites. When identifying potential fire hazards on a construction site, the two main focuses must be on fuel and heat sources (thinking back to the fire tetrahedron). These two components are the most controllable components when preventing fires on a construction site.

4.4.1 Fuel

Building materials, packaging, protective materials, chemicals, gasoline, propane, waste materials, all can contribute to the fuel load.

Below are common fuel sources for construction projects:

- Wood materials (framing, temporary construction, protection)
- Packaging (cardboard, plastics, paper)
- Wooden pallets/crates
- Insulation
- Fuel storage (flammable and combustible liquids and gases)
- Batteries for tools and other equipment
- Carpet
- Furniture
- Formwork

4.4.2 Heat

Heat, which is capable of igniting a fire, can occur from various construction processes and equipment.

Below are common heat sources that can be generated on construction projects:

- Hot work activities (including welding, cutting, thawing, soldering)
- Temporary heating systems (electric heaters, propane heaters, diesel heaters)
- Hot-applied roofing operations (kettles, heaters, torches)
- Smoking products and lighters/matches
- Cooking equipment
- Electrical equipment
- Internal combustion-powered equipment

Hot work is a general term used for any work activity that produces heat, sparks or flames. Welding, soldering, cutting and thawing are some of the more common types of hot work seen on a construction site. Hot work is often necessary for many construction tasks. Hot work brings one of the four components needed for a fire to a construction site; therefore, it is important to make sure the other required components are not present to prevent a fire. The CFSP should identify construction tasks that will require hot work and outline the proper procedures to ensure it is carried out safely. Chapter 12 will expand on hot work activities and best practices.

4.5 Mitigation Strategies to Reduce Fire Hazards

The plan will address each potential fire hazard anticipated on the project and provide a mitigation plan. The process for reducing fire risks on a project undergoes the following analysis:

1. **Can the fire risk be eliminated altogether?** Construction means and methods will be considered to see if alternative materials, processes can be used to eliminate the hazard proposed.
2. **Can the fire risk be reduced?** The next best thing after eliminating a hazard entirely is to reduce that hazard as much as possible.
3. **How can the remaining fire risk be managed to prevent a fire?** The remaining risk after the first two steps is exhausted, leaving you with what risk cannot be avoided.

Examples of this risk reduction process will be included in the forthcoming chapters.

4.6 Management of Fire Protection Systems

In existing buildings, it is essential that critical fire protection systems remain operational during construction to continue to protect occupants and the building. Modifications and impairments to the systems must be thoughtfully coordinated to ensure the level of safety in the building is not jeopardized. Protection of these fire protection systems from construction operations must also be considered.

In new buildings, priority must be given to the installation and activation of fire protection systems and other passive fire protection systems.

4.7 Impairments

An important section of the CFSP is dedicated to fire protection systems. In existing buildings, these systems are active and careful coordination is critical to ensure the systems remain active during the construction process. Chapter 14 will discuss the impairment procedures in greater depth.

4.8 Pre-Fire Plan

A pre-incident fire plan, pre-fire plan or site logistics plan is a visual tool that provides a plan of the project site with an emphasis on critical fire protection and life safety elements. This plan accompanies the CFSP to provide a visual reference. This plan is often updated periodically throughout the construction project as site conditions change. The plan identifies the location of pertinent elements that the FPPM, first-responders and other building personnel will need to know. This plan will be included in the CFSP, posted on the site, and communicated to all necessary onsite personnel.

The plan is specific to each building, and even specific to each project within the building. It outlines specific project elements such as egress routes and the location of critical life safety infrastructure.

The following items should be included in the pre-fire plan:

- Exits
- Fire hydrants
- Fire department connections for sprinkler systems
- Sprinkler control valves
- Fire alarm devices
- Utility and electrical disconnects/shut-offs
- Fire extinguishers
- Fire department access to site
- Fire command post
- Muster points
- Key boxes for fire department emergency access
- Site fence and gate access
- Construction field offices
- Dumpsters
- Flammable liquid and gas storage
- Hoists/elevators
- Crane locations

Figure 4.2 illustrates an example of a pre-fire plan and its contents.

4.9 Defined Roles and Responsibilities of Key Stakeholders

4.9.1 *Owner's Role*

The owner of the project is ultimately responsible for developing the master construction fire safety program. NFPA 241 puts the responsibility onto the owner for the creation of this plan and the delegation of a responsible person, the Fire Protection Program Manager. This master plan is building-wide to set forth building requirements, pre-fire plans, impairment procedures, hot work requirements, waste management, assignment and defining of roles and other critical information. This master plan becomes the foundation for the development of project-specific CFSPs for each ongoing project in the building. The owner must designate a primary FPPM and alternate(s).

Figure 4.3 illustrates the relationship and hierarchy amongst the various stakeholders on the project.

It is possible during ground-up, new construction, where a single general contractor (GC) controls the project, that the majority of the responsibility can

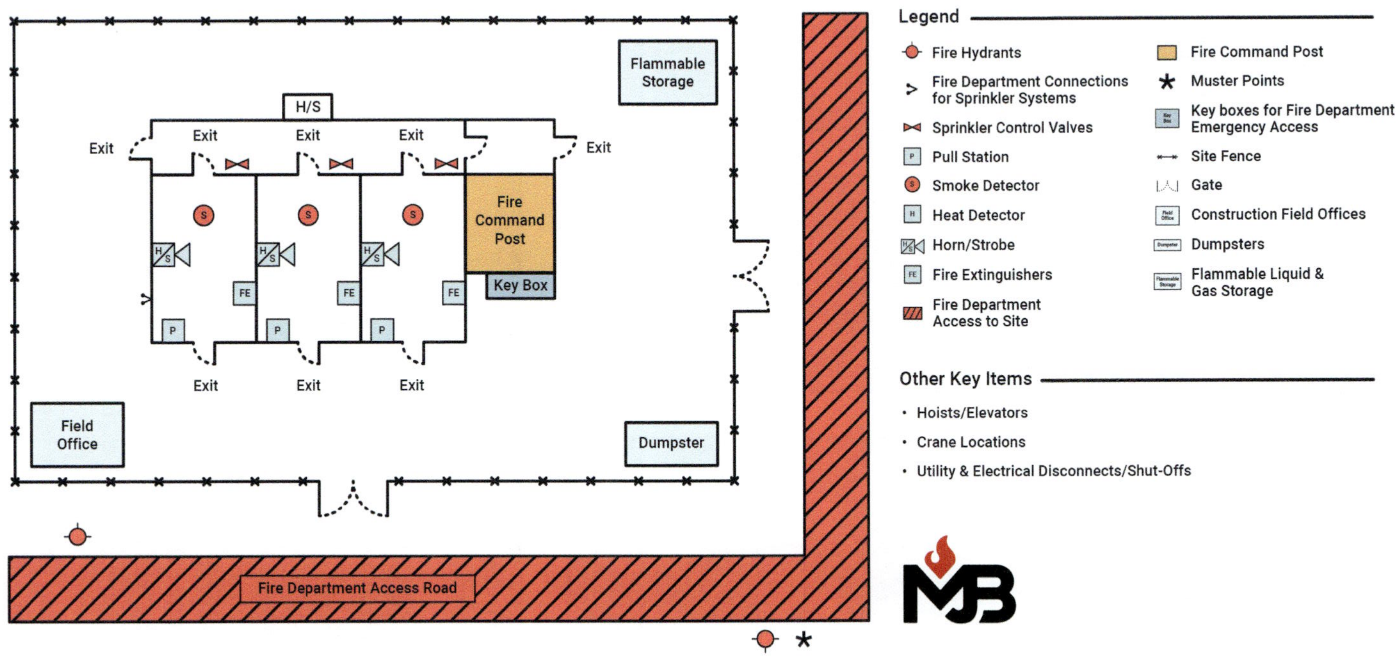

Figure 4.2 An example of a pre-fire plan and key components. Image illustration by St. Germain Design Co.

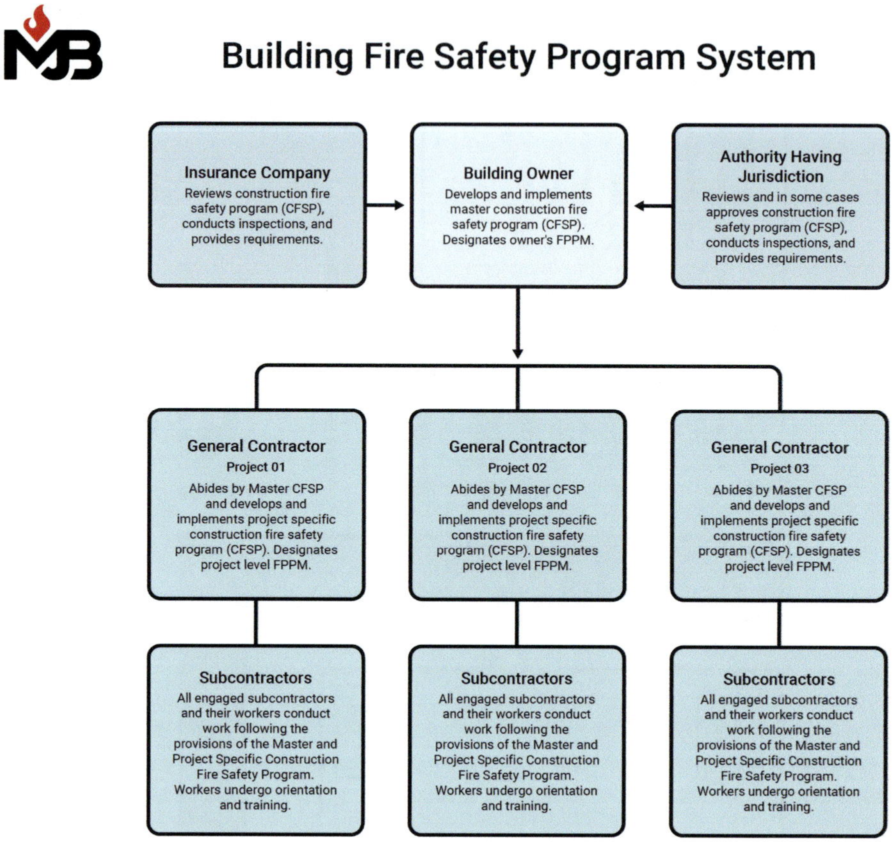

Figure 4.3 Construction Fire Safety Program Hierarchy. Image illustration by St. Germain Design Co.

be delegated to the GC. However, in existing buildings, multi-tenant, mixed-use buildings, the owner must manage the construction fire safety program and play an active role as the FPPM.

If you think of the example of a multi-level building with multiple tenants, it demonstrates why the owner must step in and manage this process. If a tenant on the first floor is under construction, they hire a general contractor, pull a permit, and start their work. Their primary focus is on their first-floor tenant space, not anywhere outside of that. They are essentially operating in a vacuum. Their construction activities may intentionally or unintentionally negatively impact the remainder of the building. Some common examples are construction materials or equipment impeding egress routes. Items placed in corridors, exit stairs and doorways can prevent free passage through these areas. If fire protection systems such as sprinkler, fire alarm and smoke control systems are impaired without close management, systems can be ineffective and provide a false sense of security. If fire protection systems are left disabled for extended periods, the construction zone will be unprotected, and the remainder of the building will be left in this vulnerable situation. These projects are often tied to tight budgets and schedules. So, when there is a hurdle in the way that might affect the budget or schedule, a decision might be made that is in the best interest of the project, but not the building.

If you add another project with another general contractor, the problem compounds. Then two projects are operating in a vacuum, presenting major risks to the building, occupants and construction personnel. This is a recipe for a disaster if not managed diligently.

The ownership of the building has a lot of skin in the game. The owner has a strong interest in maintaining the building in a safe condition for occupants and driving business to the property. The owner also fosters several relationships that are crucial for operating the building as desired. Strong relationships with the tenants, community and AHJs are critical. Unsafe construction activities, unpermitted work, nuisance fire alarms, poor notification for work impacts and testing, reflect poorly on the owner, even if these items are not directly managed by them. Therefore, it is in the best interest of the owner to insert themselves into this process to achieve desired, successful outcomes.

4.9.2 General Contractor's Role

The general contractor is typically handed the responsibility of designating a project-specific FPPM and alternate at the project level. The project-level FPPM is responsible for carrying out the day-to-day, onsite duties of the fire prevention plan. The FPPM must communicate back to the Owner's FPPM as necessary to ensure a proper channel of communication is established. The project-level FPPM is responsible for ensuring the subcontractor's and individual workers' compliance with the overall CFSP.

Each individual project in the building, typically defined as each separately permitted project, must develop a project-specific CFSP. Therefore, in a multi-tenant building, each tenant and their general contractor must develop a project-specific CFSP for review and approval by the owner, who ensures it aligns with the overall master CFSP.

4.9.3 Subcontractor & Individual Worker's Role

All subcontractors and their workers engaged in the construction project must have an understanding of the CFSP, especially as it relates to the specific trade's work. It is the responsibility of the general contractor to ensure this awareness and education take place.

For example, steelworkers must have an understanding of the hot work procedures to ensure their welding, cutting and other hot work activities align with the plan. Fire alarm and sprinkler contractors must understand the impairment procedures specific to the building. Carpenters working with wood products must have an understanding of the waste management plan to ensure combustible waste products are being gathered and discarded as soon as possible.

4.9.4 Authority Having Jurisdiction/Fire Department's Role

Authorities having jurisdiction and fire departments typically fit into the CFSP by being the group that reviews and, in some cases, approves the construction fire safety plan. The contents of the CFSP are important to the fire department as they provide critical information that plays into their response in the event of an emergency. It includes information on site access, active fire protection systems and impairments, hot work activity, emergency contacts and other critical information. Chapter 16 will review the AHJ and Fire Department relationship in detail.

4.9.5 Insurance Role

Insurance companies may not play an active role in managing the construction fire safety program, but that does not mean they do not influence the plan or have an interest in the plan. Insurance companies may have specific requirements on fire protection impairments, construction processes and construction materials that could impact the plan. Furthermore, insurance representatives may conduct routine site inspections with observation reports. Fire prevention related observations may be cited to be rectified by the project team. Insurance policies may be influenced by the quality of the CFSP in place. Where many controls are in place to prevent fires, the insurer may see the building as a lower risk compared to one without a CFSP.

4.9.6 3rd Party Consultant Role

In some cases, a qualified 3rd party can be designated by the owner or general contractor to serve as the FPPM. In scenarios where the owner and general contractor do not have a qualified person or do not have the staffing resources to designate a person into this role, a 3rd party can be introduced. This 3rd party steps in to fulfill the requirements and responsibilities of the owner or GC.

Reference

1. Sherrill v. SouthPark Charlotte Apartments LP, No. 24CV035492-590, Mecklenburg County Superior Court, electronically filed July 31, 2024.

Chapter 5

Fire Protection Program Manager (FPPM)

5.1 Fire Protection Program Manager (FPPM) Role

The Fire Protection Program Manager (FPPM) is the single most important individual for orchestrating the construction fire safety program. The role must be assigned thoughtfully to ensure the FPPM has the skills and experience to successfully serve in the role. The FPPM is also involved in the planning of fire protection systems in the building to ensure code minimums are met and designed in a manner that meets the project team's expectations.

FPPM can represent either "Fire Protection Program Manager" or "fire prevention program manager." Although there are differing opinions, these titles are generally interchangeable within the construction industry. NFPA standards use "fire prevention program manager" to signify the FPPM role.

The Fire Prevention Program Manager (FPPM) title stems from NFPA 241, *Standard for Safeguarding Construction, Alteration, and Demolition Operations*. With each update to the standard, the FPPM roles and responsibilities have expanded and become more defined. Recent editions of NFPA 241 highlight the importance of the FPPM as a key figure in the construction fire safety program.

Over the past several decades, this role has gained a lot of attention on construction sites. There has been additional pressure from authorities having jurisdiction (AHJ) to not only have this role filled on the construction site, but to have it carried out by a qualified individual.

5.2 Evolution of NFPA 241 and FPPM Role

I had the opportunity to discuss the history of NFPA 241, *Standard for Safeguarding Construction, Alteration, and Demolition Operations,* with Kevin Carr, a Senior Fire Protection Specialist at the National Fire Protection Association. Kevin currently serves as the Staff Liaison on the committee dedicated to NFPA 241 and has served in this capacity for seven years. Kevin has a wealth of knowledge on the standard and the world of construction fire safety over his fifteen years at NFPA.

To fully understand where the need for NFPA 241 and other fire and building codes stemmed from, you must understand the time in which there were leaps and

DOI: 10.1201/9781032720210-6

bounds in code advancements. There was a significant shift in urban development leading to larger and more condensed buildings in the late 19th and early 20th centuries. While the threat of a fire in a building wasn't a new phenomenon, the extent of the devastation and loss of life grew based on larger buildings being constructed and their exposure to other adjacent structures. This became the catalyst for the development of many building and fire codes. NFPA 241 was another result of this initiative to safeguard structures from fire.

Kevin Carr explained how this first came to be:

> Technically, the first formal report, Recommended Good Practice Requirements for Construction Operations, was tentatively adopted at the 1930 NFPA Annual Meeting. The 1933 edition, finalized in May of that year, was similarly titled, and received the NFPA 241 designation. It cost $0.25, and was 8-pages, cover to cover.[1]

Further modifications to the document were made in the 1950s and 1960s:

> The 1958 edition, finalized in May of that year, saw the title change to Standard for Safeguarding Building Construction Operations. While titled as a standard, the explanatory text from the document states "NFPA standards are purely advisory as far as the Association is concerned, but are widely used by law enforcing authorities in addition to their general use as guides to fire safety." This language was removed from the 1968 edition and aligns more applicably with current standard mandatory language.

An interesting part of my conversation with Kevin revolved around the role of the FPPM. This role, although not always referred to as "Fire Prevention Program Manager" or "FPPM," has been constant throughout the evolution of the standard.

Kevin described that the original 1930 and 1933 editions used the term "fire warden." He provided the definition of this role:

1933 Edition
26. Fire Warden

> A person of intelligence and tact should be appointed as a fire warden and vested with authority to supervise the installation and maintenance of the recommended fire protection appliances and fire prevention measures, the removal of all unnecessary combustible material and waste and the supervision of adequate watchman and supervisory service.

The 1980 edition revised this role as "*a capable and qualified person having the necessary authority shall be placed in charge of fire protection.*"

The term, as we know it today, fire prevention program manager or FPPM, was first introduced in the 1986 edition of NFPA 241. At that time, the technical committee performed a complete rewrite of the text. Carr shared Annex material, which provides further information and qualifications of the FPPM:

NFPA 241 – 1986 Edition

> A-5-1.1 Make one person responsible for the protection of property from fire. This person should see that the proper procedures for controlling fire hazards are set up and should have full authority to enforce them.
>
> The person should be appointed by the owner. Where an entirely new structure is being constructed, the owner should see that specifications for new buildings contain a clause stating that the "contractor will take all reasonable precautions against fire in accordance with good fire protection engineering practice.
>
> Responsibly for loss prevention rests with the owner. However, loss prevention recommendations are normally accomplished by the contractor. To assure that recommendations are carried out promptly, the owner's assistance may be needed.

This evolution of the standard, as well as the FPPM role, made two things clear. First, the FPPM plays a critical role on construction sites to manage fire prevention and implement the fire prevention program. Secondly, the owner is ultimately responsible for fire prevention in partnership with the contractor. These have remained constant as the standard has evolved since the 1980s.

5.3 FPPM Qualifications

The FPPM is required to be a qualified individual who has knowledge of fire protection systems, construction processes, fire prevention protocols, permitting and code knowledge. Similarly to how the complexity of a construction fire safety plan is directly correlated to the complexity of the construction project, the required qualifications of the FPPM have a direct association with the intricacies of the project.

The FPPM on a small, multi-family, residential house will need different qualifications and knowledge than an FPPM on a high-rise, multi-use building. The small residential building may only have smoke alarms and fire extinguishers as its life safety systems. Whereas high-rise buildings are equipped with several fire protection and life safety systems. Sprinkler systems, standpipes, smoke control systems, fire service elevators and addressable fire alarm systems are often seen in large buildings. The FPPM needs to not only understand what systems are present in the building, but they must also understand how these systems operate, how they interact with each other, and how they need to be maintained to ensure they are in good working order.

5.4 FPPM Education

Those fulfilling the duties of the FPPM must have the necessary qualifications and education to successfully carry out the required functions. The FPPM's education does not necessarily need to come from a fire protection engineering background. The examples below show different backgrounds functioning as the FPPM:

1. **Fire protection engineering education.** This background is very relevant to coursework in fire protection systems, building fire safety, building codes and fire fundamentals. Additional education may be needed to understand construction processes, permitting and general project management.
2. **Construction safety education.** This is a well-rounded education in not only construction fire safety but also a big focus on construction safety in general. There is a lot of attention on OSHA requirements.
3. **General contractor personnel.** Individuals working for general contractors gain a great amount of knowledge in construction by experiencing many projects first-hand. While the construction knowledge is very strong, additional education may be needed on fire prevention and safety.
4. **Property managers and building engineers.** The property management team manages the building post-construction. These individuals have a strong understanding of the installed mechanical, electrical, plumbing and fire protection (MEP/FP) systems in the building and recognize the details of the specific building after years of working at the property. They are involved in maintaining the equipment in the building, including life safety equipment. The property management team often encounters smaller construction projects after the main construction project is completed. Additional education may be needed to understand fire prevention safety.

Based on the skills the FPPM has and what skills they may need to strengthen, various educational resources are available to them:

Fire Prevention Education:

- Fire Protection Engineering degrees and courses
- Fire safety courses
- NFPA FPPM Training Certification
- FPPM training courses offered by industry specialists

General Construction Safety Education:

- Board of Certified Safety Professionals (BCSP)
  - Safety Trained Supervisor (STS)
- Safety and Occupational Health Degrees and courses
- OSHA – 10 Hour
- OSHA – 30 Hour
- Construction safety courses offered by industry specialists

Construction Operations Education:

- Construction Management Degrees and courses
- Construction Supervisor's License
- Construction courses offered online or locally

5.5 FPPM Building Specific Knowledge

The FPPM must familiarize themselves with the fire protection and life safety systems in the building. Relevant documentation must be located to gather information on these systems. The following items should be located and readily available for both base building and tenant improvement projects:

- Architectural Plans
- Life Safety Plans
- HVAC Plans
- Electrical Plans
- Plumbing Plans
- Fire Protection Plans
- Fire Alarm Plans
- Fire Alarm and Fire Protection Narratives
- Smoke Control Narratives
- Testing and Inspection reports of all fire protection systems
- Emergency Action Plan/Crisis Management Plan

As-built drawings and shop drawings are extremely helpful when trying to learn the systems in the building. Often, the plans developed by the engineer of record are schematic in nature and show intent, requirements, and overall compliance with applicable codes. However, they often lack the detail that would be included in the shop drawings by the installing contractors. These show details such as the precise location of valves, pipe locations, risers, and their sizes.

FPPMs should take advantage of the subject matter experts they are surrounded by. For example, the FPPM can connect with the fire sprinkler contractor or the fire protection engineer on the project to help understand how these systems function. People love talking about things they know and are passionate about, so often these types of individuals can be a great resource to educate the FPPM.

Courses and training are available on various fire protection and life safety systems. These should be explored by the FPPM in areas where they have a lack of understanding. Chapters 6 through 11 provide a general explanation of various fire protection systems and life safety elements, and how they operate. These chapters are meant to serve as a foundational understanding of the systems to allow the FPPM to perform its duties.

FPPMs must also have a thorough understanding of construction operations, processes, and schedule milestones, especially as they relate to their job function. By understanding how a construction project evolves over the course of the project, they can properly plan when various fire hazard activities are present. The FPPM must analyze the project schedule and identify areas that impact the construction fire safety plan. See example below of the various project milestones:

- Excavation and foundations
- Framing/erection of structure
- Exterior façade
- Rough mechanical/electrical/plumbing/fire protection/fire alarm
- Interior finishes
- Final MEP/FP installation
- Livening of fire protection and life safety systems
- Final inspections and acceptance testing
- Certificate of occupancy

The FPPM shall note these important project milestones to ensure updates to the construction fire safety plan are made as appropriate as site conditions change.

The FPPM must utilize a fire prevention checklist that is used for frequent inspections of the construction site. This checklist should be included as an appendix to the construction fire safety plan. The FPPM inspection checklist and inspection techniques are described in Chapter 15. The checklist is developed based on the contents and requirements found in the construction fire safety plan.

FPPMs communicate daily with owners, general contractor personnel, subcontractors, workers, and building and fire inspectors. Therefore, FPPMs must have strong communication skills. The FPPM must build strong relationships with the workers and subcontractors to get buy-in into the construction fire safety program. When a strong relationship is established, it will be possible to have more meaningful impacts on the project. If the only time the FPPM is communicating to the project team is when issues arise or for reprimanding individuals for not following the plan, there will be a lot of resistance to the FPPM and the construction fire safety plan in general. Whereas if the FPPM fosters personal relationships with these integrated parties, is part of project meetings and provides helpful recommendations, a good working relationship can be established. Therefore, when a high level of trust is established between the FPPM and other working parties, the FPPM's duties will be able to be carried out to a higher level, creating a larger impact on the project.

5.6 FPPM Primary Roles and Responsibilities

The FPPM has the following main responsibilities as part of the construction fire safety plan:

- Managing Compliance with Fire Safety Codes
- Risk Assessment and Fire Safety Planning
- Fire Protection System Management

5.6.1 Managing Compliance with Fire Safety Codes

The FPPM must familiarize themselves with all applicable codes, standards and permitting requirements in the specific location they are working in. The FPPM or the code consultant on the project must establish the basis of codes and standards on the project. When code questions arise, the number of times the incorrect code is used or the incorrect edition is applied is very surprising. It is important to understand how to correctly navigate the code to find accurate information to provide to the project team. The FPPM should have access to the most commonly used codes. It is recommended to develop a code library. These usually include the building code, fire code, fire sprinkler system standard, fire alarm standard, NFPA 241, and others that apply to the project. The list below outlines the codes that should be maintained by the FPPM:

Code/Standard	*Examples*	*Description*
Building Code	• International Building Code (IBC) and applicable amendments based on the jurisdiction of the project. • NFPA 101 – Life Safety Code • NFPA 5000 – Building Construction and Safety Code	Based on the location of the project, some combination of these codes is likely adopted. Ensure all applicable amendments and correct editions are used.
Fire Code	• International Fire Code (IFC) • NFPA 1 – Fire Code • Local city/state fire codes/ordinances	Ensure all scoped-in requirements/ standards/codes are identified in the specific "fire code" that is applicable.
Construction Fire Safety Codes	• NFPA 241 – Standard for Safeguarding Construction, Alteration, and Demolition Operations • NFPA 51B – Standard for Fire Prevention During Welding, Cutting, and Other Hot Work	The contents and requirements in these codes should be understood, as these are two of the major standards influencing a construction fire safety plan.
Fire Protection and Life Safety System Codes	• NFPA 10 – Standard for Portable Fire Extinguishers • NFPA 13 – Standard for the Installation of Sprinkler Systems • NFPA 14 – Standard for the Installation of Standpipe and Hose Systems • NFPA 70 – The National Electrical Code • NFPA 72 – National Fire Alarm and Signaling Code • NFPA 92 – Standard for Smoke Control Systems	These standards set forth design and installation requirements for the particular fire protection system.

(*Continued*)

Code/Standard	*Examples*	*Description*
Testing and Inspection Standards	• NFPA 25 – Standard for the Inspection, Testing, and Maintenance of Water-Based Fire Protection Systems • NFPA 72 – National Fire Alarm and Signaling Code	Familiarity with requirements found in these standards is important to ensure critical life safety infrastructure is being properly maintained.

Knowledge and understanding of the applicable codes and standards are a huge asset to a project. Building and fire codes are the basis of all minimum construction requirements. By understanding the codes and standards, the FPPM will be able to navigate and guide the project through difficult code challenges. The FPPM will also be able to communicate effectively with the building and fire department by confidently showing understanding of the codes.

5.6.2 *Risk Assessment and Fire Safety Planning*

Conducting complete fire hazard assessments and risk evaluations is the foundation of effective fire safety management. The process involves identifying potential fire hazard risks in a systematic approach. Next is analyzing the environmental conditions and other vulnerabilities that could be present, such as the lack of fire protection systems or the presence of fuel or ignition sources. When the risks are identified, a specific fire prevention strategy must be developed to mitigate the risk, sometimes by eliminating it entirely or reducing the risk. Fire prevention strategies may include the development and implementation of safe procedures and protocols. In other situations, fire protection systems may be utilized to overcome these risks. A comprehensive fire safety plan includes both preventive measures as well as response procedures to establish proper channels of communication in emergency situations.

5.6.3 *Fire Protection System Management*

One of the main responsibilities of the FPPM is managing the design, installation and ongoing maintenance of all fire protection and life safety systems. This involves closely working with design teams and engineers to ensure the proposed system's design will meet the building's needs. The design must adequately protect the proposed use of the building. It must also meet any aesthetic goals, operational desires and other needs of the owner and management teams. Proper planning is necessary to ensure that the fire protection systems can accommodate future expansion and modifications. The systems must be designed and installed in accordance with applicable codes and standards. The FPPM must verify that the installation meets the design intent, installed in accordance with approved plans.

Following the installation of the fire protection systems, the FPPM must ensure the systems are part of a maintenance and testing program to ensure the system remains in good working order. Monitoring the lifecycle of fire protection components is important to ensure elements are being repaired, replaced and upgraded to maintain optimal performance. A proactive approach guarantees the building's fire protection systems will work as designed. Occupants of buildings and homes depend wholeheartedly on these systems for providing adequate time for safe evacuation. It will also ensure the building stays compliant with regulations for maintaining such systems.

Reference

1. Kevin Carr (Senior Fire Protection Specialist at the National Fire Protection Association (NFPA) and Staff Liaison for NFPA 241), interview by author, March 21, 2025.

Part II

Fire Protection Systems and Life Safety Features

Chapter 6

Fire Alarm System Fundamentals

6.1 Fire Alarm System History

In the 16th, 17th and into the 18th century, as civilizations grew and became denser, the need to notify the community of fires became important. In those days, notification was mainly achieved by the ringing of bells by a fire watchman at designated fire watchtowers. Specific bell-ringing signals were developed with a various numbers of rings, intervals and pauses that would indicate what the notification was for.[1]

Fire alarm systems have seen significant advancements over the past century and a half. The first city-wide fire alarm notification system was seen in the mid-19th century in Boston, Massachusetts. This system was developed by Dr. WM F. Channing. This system consisted of fire alarm boxes located throughout the city that were connected back to a central monitoring station. Each box would send a unique signal that would correspond to a specific location within the city. The code would be translated to a specific address and communicated to the fire department. There was an official communication from Channing to the City of Boston dated March 24, 1851, outlining how this system would function. Also included were financial information, benefits, reliability information and a proposed city of Boston map which showed signaling and alarm circuits running through the city.[2] This was a huge advancement at the time, as it allowed much faster response times from the fire department (Figure 6.1).

Although present-day technology has raced past this early innovation of fire notification, the system still remains in operation in Boston – operating very closely to the plan originally laid out by Channing. This same technology exists in many other towns and cities across the United States. It serves as an independent fire notification system that can be an alternate means to notify fire officials in an emergency. This system does not rely on the internet or phone lines, so it truly is an independent system. The development of 9-1-1, cell phones and fire alarm systems with automatic calls to fire departments have certainly made these systems more obsolete, but they still exist and can be used if necessary (Figure 6.2).

In 1890, the first electronic fire alarm was invented by Francis Robbins Upton.[3] In 1902, George Andrew Darby invented a smoke detector.[4] These devices and

DOI: 10.1201/9781032720210-8

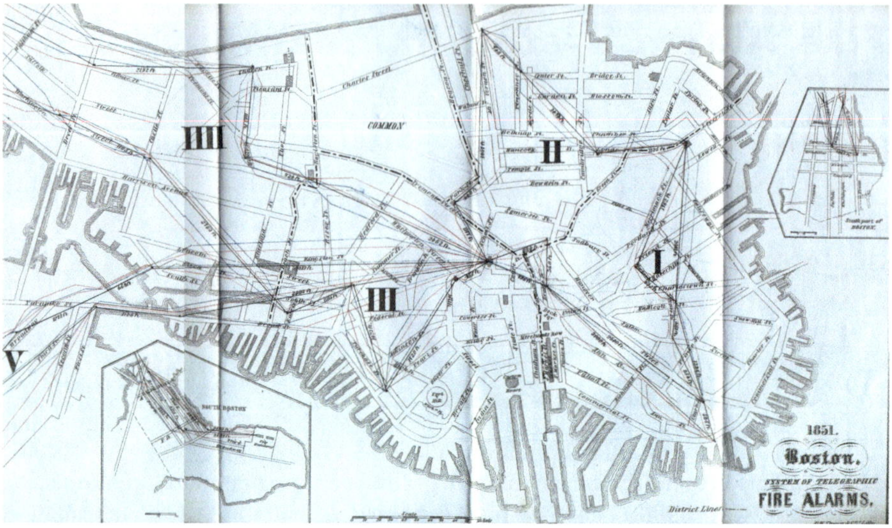

Figure 6.1 Communication from WM F. Channing on System of Fire Alarms City Diagram.

systems were basic and not widely used until the mid-20th century, when additional advancements were made.

The first fire alarm systems installed in buildings were conventional fire alarm systems. Although addressable fire alarm systems control the market today, they were not introduced until the 1980s. This chapter will describe the conventional and addressable fire alarm systems and their components.

6.2 Fire Alarm System Goals

The goal of a fire alarm system is to:

1 Detect the fire
2 Notify occupants
3 Notify the fire department
4 Enact building systems tied to fire alarm system activation

It is important that all these steps are accomplished as quickly as possible.

6.3 Conventional Fire Alarm Systems

Conventional fire alarm systems were the first type of fire alarm systems rolled out in buildings. Many conventional fire alarm systems are still in use today in existing buildings and can be installed in new, smaller buildings, although they are less common.

Figure 6.2 A present-day fire box located in Boston. Photograph by the author.

Conventional fire alarm systems have one or more zones, each with a dedicated circuit containing fire alarm devices. These zones are tied back into the fire alarm control panel. When a fire alarm device goes into alarm, a signal is transmitted back to the fire alarm control panel. However, the information and location of the alarm are limited compared to an addressable fire alarm system. The conventional fire alarm system only shows that a specific zone has an alarm. It is not possible to pinpoint the exact location of the activated device. The downside is the delay in identifying the location of an activated fire alarm device. A single zone may cover multiple tenants, floors, rooms or spaces. Therefore, the response time to find the area of the incident may be delayed compared to an addressable system.

However, in some cases, a conventional system can be more cost-effective and the installation can be simpler. Therefore, conventional fire alarm systems may be appropriate based on the specific building.

6.4 Addressable Fire Alarm Systems

Addressable fire alarm systems control the market for new fire alarm system installations presently. This is due to the many advantages these systems have over the conventional type. Addressable systems work on the basis of each fire alarm device having its own unique "address." Each fire alarm device is programmed into the fire alarm control panel system. It operates more similarly to a computer than an analog approach seen in conventional systems. Each initiating fire alarm device has a unique address in a specific format, including: NODE:MODULE:DEVICE. Such as N01:M01:D001, meaning node 1, module 1 and device 1. Furthermore, this device can have additional information associated with it, such as the location, floor level and type of device. Information such as "Level 1 – Office 111 Smoke Detector" could be included. This information would be transmitted to the fire alarm control panel when activated to assist first responders in identifying the location of the incident.

Each individual initiating device can be monitored by the fire alarm control panel. This ensures that there is sufficient connectivity to each device and each device is in good working order. If dust or debris enters a smoke detector's chamber, a warning would be sent to the fire alarm control panel. This issue could be resolved before the smoke detector either becomes defective and stops functioning properly or causes a false alarm.

The reliability of an addressable fire alarm system is also greater than the conventional type. All connected devices are continuously reporting back to the fire alarm control panel to ensure connectivity.

Addressable fire alarm systems can also include emergency voice alarm communication features. These are seen in large, high-rise, healthcare, educational and other high-occupancy buildings. Specific messages can be transmitted throughout desired areas of the building. These messages can be customized to meet the project's needs. A public address system (PA system) can also be included as a function of the addressable system. This allows first responders or fire alarm testing personnel to project messages throughout the building from the fire alarm control panel. This allows increased communication during an emergency as first responders can provide specific directions to occupants in the building.

6.5 Initiating Devices

Initiating devices are fire alarm devices that are intended to identify fire and/or smoke by automatic detection or by occupant intervention. Below are common fire alarm initiating devices found in buildings:

6.5.1 Pull Stations

Pull stations are the main method by which building occupants can manually report a fire or other emergency. They are unique in the sense that they are usually one of the only non-automatic initiating devices in the building. Whereas the rest of the initiating devices are automatic and do not rely on human intervention.

Figure 6.3 A typical fire alarm pull station. Photograph by the author.

These devices are pulled by occupants to put the fire alarm system into alarm. Pull stations often have a red box with "fire" and "pull" labeled on the device. They are typically located by exits, including main entrances and exits, and by exit stair entrances. They are strategically located along egress routes to allow occupants to pull the device while exiting the building (Figure 6.3).

6.5.2 Smoke Detector

There are many different types of smoke detectors. However, they are all designed to detect smoke produced by fire by different technologies.

- Ionization – ionization-type smoke detectors utilize a small amount of radioactive material placed between two electrically-charged plates. When smoke enters the chamber, it disrupts the current and activates the alarm. Ionization devices work best for detecting flaming fires, with minimal smoke production.
- Photoelectric – photoelectric smoke detectors utilize a small light beam within the chamber. When the light beam is interrupted by smoke, the device goes into alarm. Photoelectric smoke detectors work best for detecting smoldering or slow-growing fires (Figure 6.4).

Figure 6.4 A typical fire alarm smoke detector. Photograph by the author.

- Optical Beam – optical beam style smoke detectors utilize a light beam that is projected across an open space to a reflector and returns the light beam back to the receiver. When the light beam is obscured, the device goes into alarm. This style of smoke detection is often used in large spaces such as atriums, gymnasiums, and other large open spaces as a more efficient way of providing smoke detection.
- Air Sampling – air sampling style smoke detectors provide the earliest detection of smoke by utilizing a tubing system with ports that draw air from the protected area back to a detection chamber that analyzes the air content for specific particles that would indicate a fire. These are often installed in highly sensitive areas such as data centers, computer rooms, critical electrical rooms and other rooms containing expensive or irreplaceable contents this is to be moved to align with the type of device described.

6.5.3 Heat Detector

Heat detectors depend on the heat from a fire to activate the device. There are two main types of heat detectors:

- Rate of Rise – Rate of rise heat detectors work on the basis that if a fire occurs, there will be a significant increase in temperature over a relatively short period of time. Therefore, a specific change in temperature in a predetermined time will activate the device. Small fluctuations of temperature from normal building operations will not activate the device.
- Fixed Temperature – fixed temperature heat detectors activate when reaching a specific temperature in the heat detector's chamber. The rate of temperature increase over time is not factored into fixed temperature heat detectors. Rather,

Figure 6.5 A typical fire alarm heat detector. Photograph by the author.

a set, “fixed,” temperature is established and will only activate when that is reached. Designers will specify an appropriate temperature rating based on expected ambient temperatures of the space to reduce the chance of nuisance alarms. For example: boiler rooms, laundry rooms or other rooms that may be warmer than other areas of the building should have a higher temperature set point than normal building areas (Figure 6.5).

6.5.4 Multi-Criteria Detector

This type of detector uses multiple technologies to detect a fire and verify that the detected substance is from an actual fire and not from building operations such as steam, burnt toast, etc. These can significantly cut down on nuisance alarms in vulnerable areas.

6.5.5 Carbon Monoxide Detector

Carbon monoxide detectors rely on changes in electrical current from electrochemical sensors. The change in current correlates to a concentration of carbon monoxide molecules. Carbon monoxide detectors are often found in residential and educational buildings and in areas where carbon monoxide could be produced by building equipment.

6.5.6 *Duct Smoke Detector*

Duct smoke detectors utilize either ionization or photoelectric technology in a device specifically designed to be located within an HVAC duct. Duct smoke detectors are designed to detect smoke in moving air within the ductwork. If smoke is detected, the air handling units are designed to shut off to prevent smoke-polluted air from being circulated throughout the building.

6.5.7 *Monitor Modules*

Fire Sprinkler Flow Detector/Switch – Fire sprinkler flow detectors (or often called flow switches), are designed to detect the flow of water within fire sprinkler systems. A paddle is placed within the pipe and when there is a flow of water from an activated sprinkler head, the paddle is pushed by the water's movement in the pipe. The movement of the paddle activates the flow switch, which sends a signal to the fire alarm control panel that typically puts the system into alarm.

Fire Sprinkler Tamper Detector/Switch – a fire sprinkler tamper detector or tamper switch is used to monitor fire sprinkler system valves to ensure they remain in an open position or desired position. The tamper switch is comprised of a mechanical and electrical component that signals a supervisory signal on the fire alarm control panel if the valve is fully closed or even partially closed. This prevents the intentional or accidental closing of a valve on a sprinkler system that could restrict water flow. A sprinkler system can only operate successfully if water or other fire suppressants can get to the sprinkler heads or nozzles.

6.6 Notification Devices

Notification devices are fire alarm devices that are intended to inform building occupants of a fire or other emergency in the building. Notification devices are activated by an initiating fire alarm device or by manual intervention at the fire alarm control panel. Notification devices use audible and visual methods to notify occupants. Below are common fire alarm notification devices found in buildings:

6.6.1 *Visual Notification Only*

- Strobe only – fire alarm strobes emit a pulsing light from the unit to visually notify occupants of an emergency. The color, intensity and pulsing patterns are specified by designers for specific building applications.
- Exterior Beacon – exterior beacons are provided on the outside of the building, typically where first responders are to enter the building to gain access to the fire command center or other main entrance. These beacons are typically red and are installed to provide a clear indication of the building and response point for the fire department (Figure 6.6).

Figure 6.6 A typical fire alarm exterior beacon device. Photograph by the author.

6.6.2 *Audible Notification Only*

- Horn only – horn fire alarm devices produce a single sound tone to provide audible notification to building occupants.
- Speaker only – speaker fire alarm devices can provide sound tones and also voice signals to provide specific messages or instructions to occupants (Figure 6.7).
- Fire Bell – an exterior-mounted bell for a local waterflow alarm for the sprinkler system.

6.6.3 *Combination Notification Devices*

- Horn and Strobe – horn and strobe devices combine the horn's audible tone function and the strobe's visual light-emitting function into a single notification device.
- Speaker and Strobe – speaker and strobe devices combine the speaker's audible tone and voice signals and the strobe's visual light-emitting function into a single notification device (Figure 6.8).

Combination horn-strobe or speaker-strobe devices are often used to make the design and installation more efficient. However, strobe-only or horn or speaker-only

Figure 6.7 A typical fire alarm speaker-only device. Photograph by the author.

devices are used in areas where only one type of notification method is desired based on code requirements or the user's preference. Notification appliances come in various mounting styles such as wall-mounted or ceiling mounted to allow for some flexibility in design.

6.6.4 *Relay Devices*

Fire alarm relay devices are integrated into the fire alarm system to perform a specific function upon activation of an alarm. Relays can be tied into the building's lighting, music, HVAC, security and other building systems. The purpose of the relay is to put building systems into a specific condition upon an alarm. Lighting systems in dimly lit spaces, such as nightclubs and movie theaters, must return to a fully illuminated state upon a fire alarm to assist occupants with navigating to building exits. Audio from music, movies or other sound-producing equipment must pause to allow occupants to hear signals from the fire alarm system. HVAC equipment is sometimes tied into a fire alarm relay to shut off or return to a specific operating condition to allow other building life safety systems to operate as designed. For example, stair pressurization systems may depend on the building's HVAC system to be disabled to achieve the minimum pressure differential between the stair enclosure and the adjacent building. Various relays can also be provided

Figure 6.8 A typical fire alarm speaker and strobe combination device. Photograph by the author.

for security and other access control systems. This allows turnstiles, access control doors or spaces to release to allow for free egress.

6.6.5 *Common Misconception: Smoke Alarm vs Smoke Detector*

A smoke alarm and a smoke detector are often assumed to be the same thing. However, there is a significant difference between the two. A smoke detector is a fire alarm device that detects smoke; however, it is connected to an overall fire alarm system. When a smoke detector is activated, it puts the fire alarm system into alarm and notifies the building through the notification system (horns, speaker and strobes). Whereas a smoke alarm is a device that can detect fire and also emit an alarm signal from the same device. Smoke alarms are typically installed in residential homes.

6.7 Fire Alarm Signals

Below are the main signals on a fire alarm control panel:

Alarm – an alarm signal is generated by the activation of an initiating fire alarm device. An alarm signal will automatically activate the notification devices of the fire alarm system and will typically send a signal to the fire alarm monitoring company, resulting in a prompt response from the fire department.

Supervisory – a supervisory signal is generated when a specific initiating devices are designed to only produce a lesser signal to alert of a condition that may require intervention, but not an activation of the notification system or a response from the fire department.

Trouble – a trouble signal is generated when the monitored fire alarm circuits and devices detect an issue. Open circuits, ground faults and defective devices can produce troubles. If fire alarm wires are cut or nipped by fasteners, fire alarm devices become dislodged from their mounts, or debris enters the device, a trouble signal can be generated.

Other/Disabled – impairments require the disabling of fire alarm points to allow certain systems to be deactivated while modifications are made to the system. These disabled fire alarm points are typically listed on the panel as such.

6.8 Fire Alarm Sequence of Operations

When a signal from an initiating device comes into the fire alarm control panel, the reaction of the system is outlined in the established fire alarm sequence of operations. This specified sequence of operation from the designer indicates the result of each initiating device.

For example, the activation of a smoke detector, pull station or sprinkler flow switch will all put the system into a general alarm state and, in turn, activate the notification devices and initiate the evacuation sequence. Whereas, for example, a sprinkler tamper switch may only result in a supervisory signal on the fire alarm control panel and will not cause notification devices to activate. If a speaker/strobe device is removed from its base mounting component, this will create a trouble signal on the fire alarm control panel and will not cause notification devices to activate.

In high-rise buildings, phased evacuation procedures are often utilized. These large buildings contain many occupants, so those occupants in closest proximity to the incident are given priority for evacuation. A common sequence is the evacuation of the floor of the incident and the floors above and below. This approach evacuates those occupants in closest proximity to the event. This prevents overcrowding exits in the building to allow for sufficient egress. Further evacuation instructions can come from the fire department's manual evacuation of the building from the main fire alarm control panel. Floors outside of the evacuation zone will receive a pre-alert message throughout the fire alarm systems, speakers that will indicate an emergency in the building and to await further directions. This gives occupants early warning that there is something happening in the building and to be aware, but not to immediately evacuate.

In non-high-rise buildings, typically a general fire alarm signal will signal the notification devices throughout and evacuate the entirety of the building.

6.9 Fire Alarm Planning

The owner, property management team and FPPM should all be involved in the planning of fire alarm systems within the building. The fire alarm system must be designed with life safety as the number one priority. Secondly, the system must be designed to accommodate the specific uses of the building without being a hindrance. Improper planning can result in nuisance alarms, insufficient capacity to add additional fire alarm devices, zone isolation issues and access issues preventing proper maintenance of devices. The design and installation of a code-compliant fire alarm system may seem sufficient; however, there are many more design elements that make the fire alarm system suitable for the project's needs.

In large, mixed-use, multi-tenant buildings, fire alarm system circuits and zones should be segregated by floor levels, tenants and various uses. While all the zones will still be tied into the building's overall fire alarm system, it will create separation between them that will provide long-term benefits. Mixed-use retail areas often contain several different tenants. Each tenant space should be designed to have a fire alarm terminal cabinet within each space. This FATC allows each tenant to add their new fire alarm devices from their buildout without having to make connections outside of their space or within the base building fire alarm equipment. The FATC also allows for easier troubleshooting if there is an issue down the road with the system. Specific circuits can be removed and tested to identify the source of the issue.

When thought is put into isolating each tenant, floor and use of the building from a fire alarm standpoint, it provides additional benefits. Bypass functions can be added to the fire alarm control panel to limit the areas where fire alarm signals will be sounded. This is extremely useful in multi-tenant buildings. The example below shows the benefit of isolation and bypasses.

6.9.1 Example #1

Fire alarm testing must be performed in a 10-story mixed-use building. The building contains mixed retail on the first floor, and floors 2–10 are office space. Fire alarm testing is occurring in the office portion of the building in the early morning before office workers arrive (6 a.m. – 9 a.m.); however, the first floor contains a coffee shop tenant and a daycare tenant. These particular first-floor tenants must operate in this important morning timeframe. If fire alarm signals can be bypassed from their spaces, it will provide a much better experience for the tenants and for patrons. The bypass feature allows the fire alarm testing to occur while limiting the disturbance to the building.

6.10 Interconnected Buildings

Interconnected buildings will often have integrated fire alarm systems. This integration is necessary to have the systems communicate with each other to evacuate exposed areas. However, without careful planning, the interconnectivity could become burdensome. The example below shows the importance of careful planning in interconnected buildings.

6.10.1 *Example #2*

A sub-grade parking garage extends below two buildings above grade. Above grade, these buildings are standalone, independent structures. However, below grade, they are open to each other across the parking areas. Alarms in the parking garage should notify both buildings since there could be an immediate danger to occupants in each building. This allows stair pressurization systems to activate, which is appropriate for stairs extending down into the garage that serve each building. However, it is probably not necessary that an alarm within the above-grade portion of one building would result in putting the other building into alarm. Especially with the utilization of fire-rated floor assemblies and the parking garage as a buffer. Therefore, it is important to understand the proposed fire alarm system's sequence of operation early in the design to ensure it will not be a hindrance to the building's operation. It is much easier to modify when the building is in design than after it is already designed, permitted, constructed and accepted by the fire department.

6.10.2 *Planning Considerations*

Often, buildings are constructed initially as core and shell buildings, requiring additional design and construction to complete them to an occupiable state. When this is done, fire alarm systems are provided; however, they are provided minimally to meet the code minimum requirements. Therefore, it is critical that the future expansion of the system is considered based on the possible tenant use. This will allow a future tenant's fire alarm devices to be added to the existing building's fire alarm system without significant modifications or upgrades. Fire alarm nodes, terminal cabinets and other equipment must be sized based on the future completed fire alarm tie-ins and not the interim core and shell condition.

6.11 Review of Tenant Plans

Review of the tenant's fire alarm plans should be carried out to ensure the proposed design meets the building's expectations. The following parameters should be studied when reviewing fire alarm plans:

1 Impairment requirements – the impairment guidelines should be outlined to allow the upfront understanding of building requirements. Required notification and impairment coordination meeting obligations should be stated.
2 Fire alarm programming and tie-in – often a single fire alarm service company and programmer is the sole party performing the programming and tie-in of the new fire alarm devices. Contact information and building requirements must be outlined to ensure the expectation is clear.

3 Fire Alarm Terminal Cabinet – Ensure all interfaces to the base building fire alarm system are made at the dedicated FATC. New and existing fire alarm devices within the tenant space should be connected to the FATC. Some tenants or designers might take the approach of designing the fire alarm system as an independent, standalone system. However, it is important that the system is integrated into the base building system to ensure all areas of the building are protected.
4 Beacon – some jurisdictions may require the installation of a fire alarm beacon at the entrance of each tenant space. The specific city or town requirements or building standards should be communicated.
5 Wiring – fire alarm wiring requirements following building standards for color coding, conduit requirements and other ratings should be identified.
6 Smoke Detector Problematic Areas – smoke detectors in close proximity to cooking operations or cleaning operations could pose a risk of nuisance alarms from smoke, steam and other byproducts. Fire alarm devices must be listed for the specific application where they are installed, such as exterior locations, walk-in coolers, high moisture areas and other areas subject to environmental factors. Heat detectors should alternatively be used to prevent nuisance alarms in some applications.
7 Carbon Monoxide – CO detectors are often provided in kitchen areas of restaurants and similar uses. The sequence of operation for these CO detection systems must be reviewed. The CO detection system is often driven by a plumbing and gas code requirement to shut down the gas solenoid valve and provide a local alarm to occupants. Integration with the base building fire alarm system may not be required by code. Operational issues can arise from supervisory and alarms stemming from power loss in the tenant space or other maintenance of the kitchen equipment. A local alarm/integration may be more appropriate.

6.12 Fire Alarm Plan Tiers

In a controlled construction arrangement, it is typical that there are three tiers or sets of documents produced for the fire alarm system. Often, fire sprinkler systems follow a similar progression of documents.

Tier I plans are plans produced by the engineer of record that establish the design basis for the project. Applicable codes and standards, design methodologies, spacing and other design parameters are set. Fire alarm devices are shown schematically on plans, without showing details of the system such as circuitry, wiring details, routing and battery calculations. These plans are incorporated into the construction set of documents that is used for permitting, bidding and contractual obligations with each subcontractor. These plans are used for a building permit.

Tier II plans, often referred to as "shop drawings", are produced by the installing contractor to provide greater detail into the system arrangement being installed. These plans are influenced by the tier I plans and must follow all established requirements. However, these plans will refine the design and connect the dots. Specific circuitry and wiring details, riser diagrams, voltage drop calculations and battery calculations are included. These plans are submitted to the engineer of record for review in conformance with the tier I documents and all applicable codes and standards. These plans are used for permits associated with the fire department, such as a fire department fire alarm permit.

Tier III plans, often referred to as as-builts, are produced by the installing contractor to indicate the final installed condition, inclusive of any field modifications or other changes. This provides the project team and owner with a conformed set of drawings that reflects the final arrangement of the system. These are extremely useful for property management, engineers and other teams involved with the maintenance of such systems.

6.13 Design-Build Approach

In some instances, it may be more efficient to take a design-build approach. This involves engaging the fire alarm company to perform both the design of the system and the installation, without the oversight of an engineer of record. This is sometimes referred to as a delegated design. The design team's drawings and specifications may not include the fire alarm and sprinkler system. The system would be designed when a fire alarm or sprinkler company is engaged.

There are many benefits to the design-build approach, such as efficiency with time and money. However, you lose the checks and balances approach that is provided in a controlled construction process. Therefore, the owner and design team should collectively decide what route is more appropriate for a given project.

6.14 Maintenance of System and Testing, and Inspection

Fire alarm systems play a crucial role in the notification of building occupants during emergencies. Therefore, it is important that they are properly maintained to ensure they remain in good working order. NFPA Standards and insurance companies are typically the main drivers of setting out minimum requirements for maintaining fire protection systems. The FPPM and property management team must understand the applicable minimum requirements and develop a program that meets or exceeds them. Proper maintenance ensures that the systems will operate as intended, prolong the lifespan of these systems, and also keep the building in compliance with codes.

Fire alarm systems installed in accordance with NFPA 72, *National Fire Alarm and Signaling Code*, must meet the minimum inspection, testing and maintenance requirements found in this standard. The majority of system components are subject to annual visual inspections and testing. Since this requires a complete test of all components, in large buildings, this is sometimes broken out into quarterly testing. Twenty-five percent of the devices are tested each quarter, meeting the 100% requirement each year. This also aligns well with the quarterly fire sprinkler testing of the waterflow devices. These are often combined since you need a fire alarm service technician on-site anyway for the testing to take place.

A clear fire alarm control panel with no trouble signals or supervisory signals must be provided at all times. If a trouble or supervisory signal is received on the fire alarm control panel, it must be promptly investigated to understand the cause. If deficient items are identified, they must be recorded and addressed in a timely manner. Defective items can interfere with the overall system's performance, compromising proper detection or notification. Trouble or supervisory signals on the system can also prevent any future fire alarm acceptance tests of the system by the authorities having jurisdiction (AHJ). Therefore, it is critical that the fire alarm system is closely monitored and any issues are quickly remedied.

Inspection, testing and maintenance records must be organized and maintained. Records could be requested by AHJs, insurance companies, prospective tenants or buyers. Records are often requested when applying for permits or when going through final inspections and sign-offs.

References

1. Harrison, Tom. "New York's Last Fire Watchtower Has Been Restored." *Smithsonian Magazine*, October 1, 2019. https://www.smithsonianmag.com/smart-news/new-yorks-last-fire-watchtower-has-been-restored-180973419/.
2. Channing, WM. F. *System of Fire Alarms in Boston, 1851*. Document No. 20. Boston Fire History. Last modified December 2019. https://bostonfirehistory.org/wp-content/uploads/sites/51/2019/12/systemoffirealarms1851.pdf.
3. Upton, Francis R., and Fernando J. Dibble. *U.S. Patent No. 436961A*, filed September 2, 1890, and issued December 2, 1890.
4. Darby, George Andrew. *U.S. Patent No. 702,122*, filed November 6, 1901, and issued March 18, 1902.

Chapter 7

Fire Sprinkler System Fundamentals

7.1 Fire Sprinkler System History

The first modern fire sprinklers were introduced in 1812 by inventor William Congreve. They were installed in the Theatre Royal in London, England. This initial style involved a storage reservoir holding water and manually operated valves to release water in a fire. Perforated piping installed in the building would fill with water and create small sprays onto the fire. These were the most popular types of fire suppression systems for the better part of the 19th century.[1] These systems provided a huge improvement in protecting buildings from fire and increased the level of life safety. Owners of buildings who suffered fires, or had large insurance premiums, or expensive contents being stored, would sometimes utilize these systems.

While these perforated systems provided many benefits, they also had many drawbacks. These systems delivered water not only where the fire was, but also everywhere else where the piping system was installed. Substantial water damage was common. Also, because of this, it was possible for the fire area to get low residual water flow and pressure, as water was also leaving the piping system in areas outside the fire zone. Corrosion and rusting were also common on the system, sometimes preventing water from flowing out. Or the perforated holes in the piping could become plugged by dirt or debris. Another substantial disadvantage was that this system depended on human intervention to detect a fire and manually operate a valve to allow the system to operate. Therefore, the system was rendered useless if the building was left unsupervised.

In the 1870s, advancements were made to allow the fire sprinklers to be "automatic," by introducing a fusible element in the sprinkler head that would release upon exposure to heat from a fire. Henry Parmelee of Connecticut created the first fire sprinkler that used solder that would soften in a fire to open holes that are sealed under normal conditions.[2] This concept was improved by Frederick Grinnell, who created a fusible link that held back a seated valve. Water in the piping system was held back by the valve and released when the fusible link was exposed to heat from a fire. This allowed the system to operate automatically and to only activate the sprinkler heads in close proximity to the fire.[3] These two advancements were revolutionary (Figure 7.1).

DOI: 10.1201/9781032720210-9

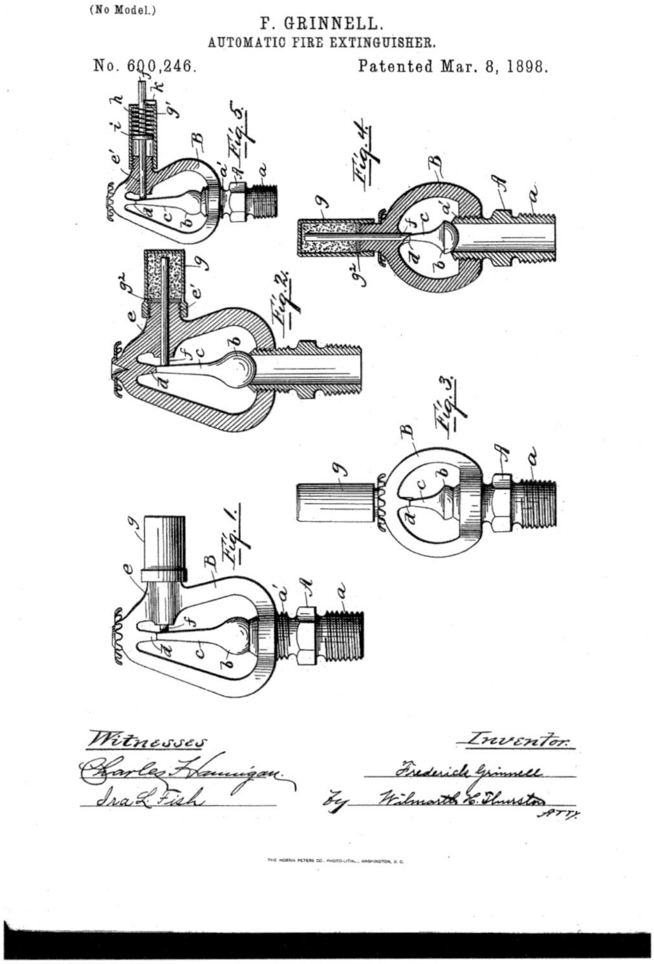

Figure 7.1 Frederick Grinnell Automatic Fire Extinguisher Patent, 1898.

With the installation of fire sprinkler systems catching on in many large warehouses, factories, manufacturing buildings, and other vulnerable buildings, the need to standardize fire sprinkler installation requirements was necessary. The National Fire Protection Association (NFPA) was founded in 1896 in Massachusetts to help unify the industry. NFPA 13 – Standard for the Installation of Sprinkler Systems was published to create a standard that could be widely adopted to create consistency in the installation and enforcement of these systems.

In the early 20th century, fire sprinklers continued to be refined. Fusible glass bulbs were introduced, replacing the metallic soldered links in some models. Fire sprinkler systems found their way into more buildings during the 20th century. Originally only required in large commercial buildings, it then expanded to many building types, sizes and uses. In today's age, it is easier to describe the buildings where sprinklers are not required than the type of buildings where they are required. In many jurisdictions across the United States, the only buildings exempt from fire sprinkler requirements are one- and two-family houses. Some jurisdictions are even requiring one- and two-family houses to have sprinkler protection.

Current-day fire sprinkler systems are robust, engineered systems. However, they operate using the same fundamental principles that were established in centuries past. Fire sprinkler systems automatically respond to a fire to provide early suppression to contain the fire. The early and automatic feature of the system is what makes it so important. The probability of total extinguishment or containment are significantly improved when a fire is caught in the early stages. Fire sprinklers provide occupants with one of the most crucial life-protecting systems by providing additional time for a safe evacuation. They also provide a great deal of property protection, preventing monumental losses.

7.2 Fire Sprinkler System Components

The purpose of a fire suppression system is to detect and suppress fires in their early stages, minimizing damage and enhancing the safety of building occupants. There are several different types of fire sprinkler systems, including wet, dry, pre-action, deluge, foam, dry agent and other types of sprinkler systems. These systems will be explored in this chapter. Their applications, purposes, and differences will be outlined. Although these systems all have slightly different functionalities, the fundamentals of the systems share many similarities. The following sections will explain the major components of the system.

7.2.1 *Water Supply and Entrance to Building*

The fire protection system's water supply is most commonly an independent connection to the municipality's water supply system. Where access to a sufficient water supply is not available or where the water supply cannot hydraulically support the designed sprinkler system, alternative water supplies can be provided, such as onsite water storage tanks or ponds that serve as fire ponds.

In some large, high-rise, or other critical-type buildings, a redundant fire protection supply must be provided to the building. By having two remote fire protection supply lines into the building, the reliability of the system is significantly improved.

In smaller residential buildings, it is sometimes permitted to have the domestic water service into the building also serve the fire sprinkler system. In these cases, it must be evaluated to ensure that the plumbing demand does not hinder the required demand of the fire sprinkler system.

The fire protection water supply line typically enters the building below grade in a basement or sub-grade level. The supply is often run directly into a fire sprinkler room in small to mid-size buildings. In large or high-rise buildings, the line may run into a water meter/supply room, then directly to a fire pump room. In either case, the supply line is provided with a double check valve assembly (DCVA). The DCVA prevents any backflow in the system. This is typical with any connection to public water supplies to prevent flow from the building back into the water supply system.

7.2.2 Fire Sprinkler Room

The fire protection supply usually runs into a dedicated fire sprinkler room. This room is a bridge between the system supply and the building's fire sprinkler system. This room contains valves and components for isolating the system and draining the system. It also contains elements to allow the testing of the system. Gauges are provided on the system to monitor pressure within the system. Waterflow and tamper switches are provided. Fire department connection piping is piped into the system.

7.2.3 Fire Pump Room

In large and tall buildings, a fire pump is often provided. The engineer will determine if a fire pump is necessary based on the hydraulic demand of the designed sprinkler system. The fire pump boosts the pressure and flow to meet the system's demand. This is usually housed in a dedicated fire pump room that is separated by fire-rated construction. In buildings with fire pumps, often this room also contains components that would be found in a typical fire sprinkler room, such as valves, testing components and gauges. There are two main types of fire pumps:

7.2.3.1 Positive Displacement Pump

These types of pumps work on the premise of compressing a fluid in a fixed volume repeatedly. There are reciprocating style systems where pistons, plungers or diaphragms are used to compress the fluid within a cylinder. There are also rotary style positive displacement pumps where gears or cogs are used to compress the fluid. These types of fire pumps have limited applications since, although high pressures can be achieved, the flow rates are much less than what can be seen in centrifugal pumps. These types of pumps can also create a pulsating effect due to the consistency of the compression and suction motion of the piston.

7.2.3.2 Centrifugal Pump

These types of pumps contain an impeller within a casing. The fluid enters the compartment with the spinning impeller at the axis. The fluid experiences the effects of centrifugal forces and is forced outward and directed to an outlet connection. These

Figure 7.2 A typical fire pump room arrangement. Photograph by the author.

style pumps can achieve high pressures and flow rates, making them very suitable for the application of a fire pump.

Centrifugal fire pumps come in different configurations, such as horizontal split case pump, vertical split case pump, in-line pumps and others. Engineers will specify a specific type based on building-specific considerations. Factors such as target pressure and flow, cost and the fire pump room size help determine the appropriate pump for the project. Horizontal split case pumps are the most prevalent as they can provide sufficient pressure and flow rates and they are economical and reliable. Fire pumps are primarily powered by either electric or diesel engines (Figure 7.2).

7.2.3.3 Jockey Pump

A jockey pump is a smaller, supplemental pump that maintains a constant pressure within the fire protection system. Large buildings have a significant amount of sprinkler piping over multiple floors and across many square feet of area. These large piped systems can experience pressure variations. It is important that a constant pressure is maintained within the sprinkler system so that the loss of pressure from the activation of a sprinkler head is quickly discovered. The quick discovery of an activated sprinkler head will be detected by the waterflow alarm and send a signal to the fire alarm system. The fire pump starts upon the pressure drop.

7.2.4 *Sprinkler Riser*

The sprinkler riser is how the fire protection system is distributed vertically throughout the building. The sprinkler riser in many cases may also serve as a standpipe system within the exit stairs, called a combination sprinkler standpipe riser. The riser spans the full height of the building. On each floor, there is a floor control valve assembly (FCVA) to individually feed each floor. Each floor level or every other floor level may also be provided with a hose connection if the riser also serves as a standpipe (Figure 7.3).

7.2.5 *Sprinkler Line Distribution (Mains and Branch Lines)*

The sprinkler main piping starts from the floor control valve assembly (FCVA) on each floor and provides water to the branch lines on the floor. The main piping is usually smaller than the sprinkler riser but larger than the branch line piping.

The sprinkler branch lines are fed from the sprinkler mains. The branch lines are provided off the main in a layout that meets the minimum spacing requirements. Sprinklers are provided off the branch lines by an outlet in the branch line piping.

In an upright sprinkler arrangement, a sprig, or a vertical short pipe, is connected to the branch line from an outlet extending up (Figure 7.4).

Figure 7.3 A typical combination sprinkler standpipe riser. Photograph by the author.

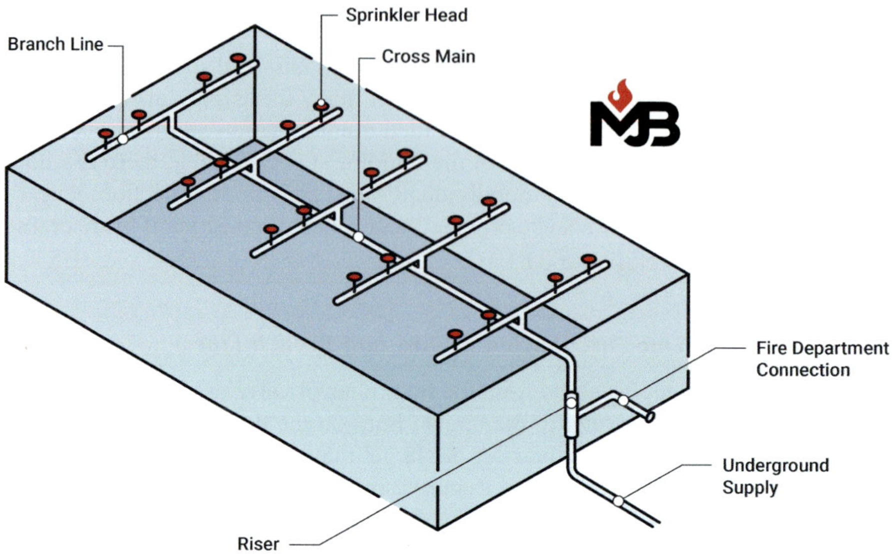

Figure 7.4 Diagram of Sprinkler Main and Branch Lines. Image illustration by St. Germain Design Co.

In pendent sprinkler systems, an arm-over, consisting of a riser nipple off of an outlet from the branch line, an arm-over horizontally, then a downward drop to feed the sprinkler.

A third, and less common, orientation of sprinklers is sidewall sprinklers. In these configurations, the branch lines run along a wall, feed sidewall sprinklers located directly below the ceiling. These provide coverage extending from the wall, different than upright or pendent, where the sprinkler water is delivered from above.

7.3 Types of Sprinklers

7.3.1 Upright

Upright sprinklers are oriented vertically with the deflector at the top, closest to the ceiling. The water is discharged from the orifice initially upwards before contacting the deflector. The deflector is specifically designed to redirect the water downwards in a certain spray pattern; therefore, an upright sprinkler should only be used in upright arrangements (Figure 7.5).

7.3.2 Pendent

Pendent sprinklers are oriented vertically; however, the deflector is at the bottom, closest to the floor. The water is discharged from the orifice downwards towards the deflector and ground. The deflector in this case is designed to create the desired spray pattern and to allow water flow downwards.

Figure 7.5 An upright sprinkler head. Photograph by Trinidade. *Fire Sprinkler*. February 27, 2013. Licensed under Creative Commons Attribution-Share Alike 3.0 (CC BY-SA 3.0). License details: https://creativecommons.org/licenses/by-sa/3.0/.

Within the pendent orientation, there are different types of arrangements:

7.3.2.1 Recessed Pendent

The piping and part of the sprinkler are concealed within the ceiling assembly. The sprinkler passes through the ceiling with an escutcheon surrounding it. This

Figure 7.6 A recessed pendent sprinkler head. Photograph by Brandon Leon, July 12, 2010. Licensed under CC BY-SA 2.0 (see https://creativecommons.org/licenses/by-sa/2.0/).

provides a more subtle look than seeing all of the components of the fire sprinkler system (Figure 7.6).

7.3.2.2 Traditional Pendent

The sprinkler and the piping are not concealed (Figure 7.7).

7.3.2.3 Concealed Pendent

All piping and the sprinkler are concealed within the ceiling assembly. The only visible element of the sprinkler is the circular, flush-mount, cover plate. This provides the most hidden option (Figure 7.8).

7.3.3 Sidewall

The sidewall sprinkler is a sprinkler designed specifically for a horizontal orientation. A sidewall sprinkler discharges water outward horizontally in lieu of providing water from above (Figure 7.9).

Figure 7.7 A traditional pendent sprinkler head. Photograph by author.

7.3.4 *Sprinkler Components*

Upright, pendent and sidewall sprinklers, although having different installation orientations, the components on the sprinkler are similar. Figure 7.10 shows the various components of the sprinkler.

7.3.5 *K-Factor*

Fire sprinklers are sized using a k-factor. Smaller k-factor sprinklers have smaller orifice sizes, whereas larger k-factor sprinklers have larger orifice sizes and discharge more water. The k-factor is specified by the designer based on the desired design density of water to be delivered to the building. For example, light hazard occupancies often have 5.6 k-factor sprinklers to deliver the code-required

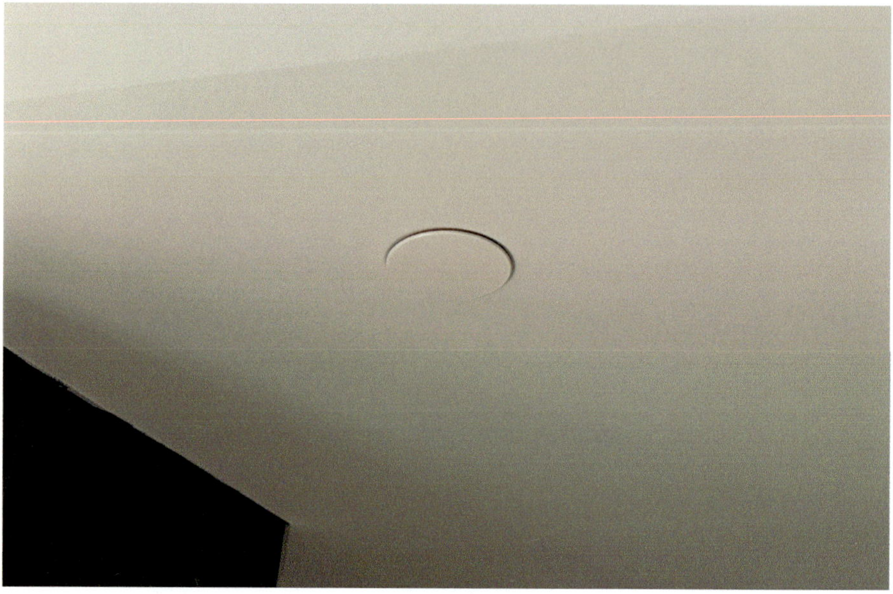

Figure 7.8 A concealed pendent sprinkler head. Photograph by author.

Figure 7.9 A sidewall sprinkler head. Photograph by Trinidade. *Wall Sprinkler*. Licensed under Creative Commons Attribution 2.0 (CC BY 2.0). License details: https://creativecommons.org/licenses/by/2.0/.

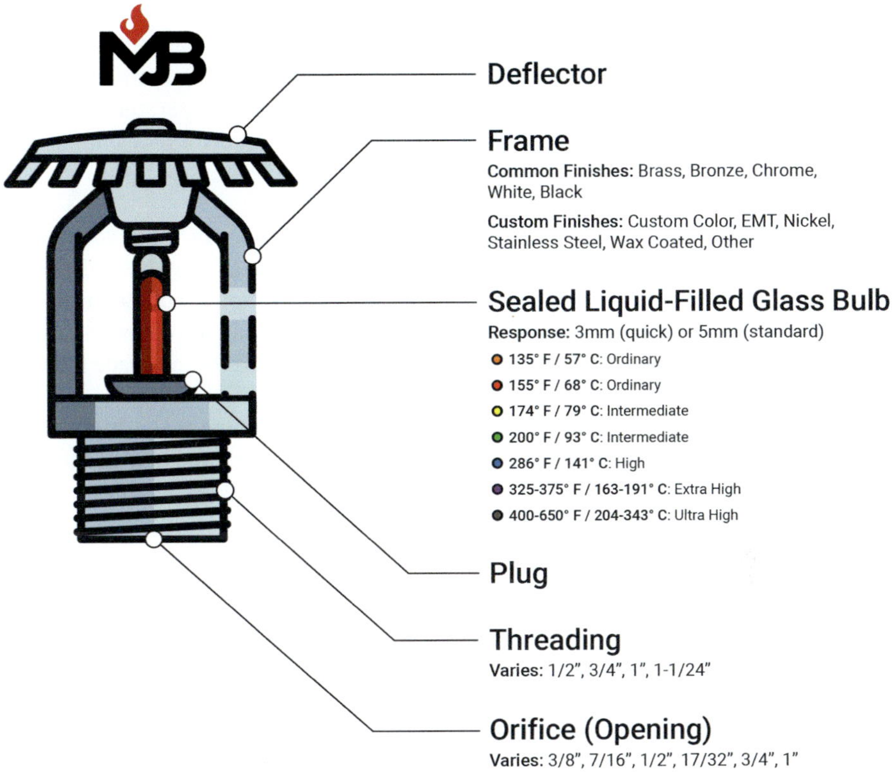

Figure 7.10 Typical Fire Sprinkler Components. Image illustration by St. Germain Design Co.

0.1 GPM/SF design density to the design area. Whereas higher k-factors, such as 14 or 25.2, are used in higher hazard occupancies where higher concentrations of water are needed.

K-factor: the sprinkler's k-factor is based on the relationship between flow and pressure:

$K = Q/\sqrt{P}$

K = K-Factor

Q = Flow (gallons per minute)

P = Pressure (pounds per square inch)

7.3.6 Temperature Rating

Sprinklers use a glass bulb or a fusible alloy element to hold the plug in place. The plug is seated in the sprinkler orifice and creates a seal to hold back the water from the fire sprinkler system. When the glass bulb or fusible alloy element reaches its temperature rating, it will break, allowing for the plug to be pushed out by the flow of water. This function is what makes the sprinkler system "automatic." Glass bulbs are most commonly used in standard sprinklers. The temperature rating of sprinklers is specified based on the ambient ceiling temperature expected. By correctly specifying an appropriate temperature rating, it ensures the sprinkler will operate as designed without unwanted activation.

For example, if a sprinkler is provided in a laundry room or other room with equipment that produces heat, the ambient ceiling temperature may be higher than a standard interior space. Therefore, to prevent the unwanted activation of the sprinkler, an appropriate sprinkler temperature rating must be provided. A higher temperature rating will allow the higher ambient temperatures of the space to occur without the activation of the sprinkler.

Conversely, it is important not to provide a temperature rating that is higher than what is necessary. If a high temperature rating is provided for a space with standard ambient ceiling temperatures, it could cause a delay in sprinkler activation as the temperatures need to increase before the rating is met.

Table 7.1 shows ambient ceiling temperatures and the recommended temperature rating and the corresponding color code.

7.3.7 *Thermal Sensitivity*

The thermal sensitivity of a sprinkler describes the response time for the sprinkler to operate. The temperature rating is the temperature that the glass bulb or fusible link will operate; however, the thermal sensitivity identifies how fast it will operate when exposed to those temperatures. This is most commonly expressed as response time index (RTI) and broken into two main categories: standard response and fast response sprinklers. These different sprinkler

Table 7.1 NFPA 13: Temperature Ratings, Classifications, and Color Codings

Maximum Ceiling Temperature		*Temperature Rating*		*Temperature Classification*	*Color Code*	*Glass Bulb Colors*
°F	*°C*	*°F*	*°C*			
100	38	135–170	57–77	Ordinary	Uncolored or black	Orange or red
150	66	175–225	79–107	Intermediate	White	Yellow or green
225	107	250–300	121–149	High	Blue	Blue
300	149	325–375	163–191	Extra high	Red	Purple
375	191	400–475	204–246	Very extra high	Green	Black
475	246	500–575	260–302	Ultra high	Orange	Black
625	329	650	343	Ultra high	Orange	Black

thermal-sensitive types are specified by the designer based on building-specific parameters and fire protection goals.

7.3.8 *Coatings*

Coatings can be applied to the sprinklers by the manufacturers to ensure the durability of the product. Sprinklers can be installed in many different applications with different environmental conditions. Environments such as marine areas, humid areas, manufacturing areas and unconditioned areas may promote corrosion on sprinklers. The corrosion over time can interfere with the sprinkler operation by delaying activation or obstructing the spray.

Coatings not performed or approved by the manufacturer are not permitted to be made to the sprinklers. Coatings that are applied to sprinklers undergo tests to ensure sprinkler functionality is not compromised. Therefore, aftermarket coatings or painting of sprinklers are not permitted. Special attention should be paid when painting occurs near sprinkler systems. Protection should be provided to prevent sprinkler heads from getting any paint on them.

7.4 Types of Sprinkler Systems

7.4.1 *Wet Sprinkler Systems*

Wet sprinkler systems are the most common type of system used in buildings today. These types of systems are pressurized with water throughout the system, with sprinkler head fusible bulbs holding back the flow of water. The activation of a sprinkler occurs when the sprinkler's fusible link is exposed to heat from a fire or other source and reaches its designed temperature. The link breaks, the plug holding back the water flow releases, allowing for the flow of water.

7.4.2 *Dry Sprinkler Systems*

Dry sprinkler systems are used in areas of buildings or structures that are not conditioned, where a minimum ambient temperature of 40 degrees Fahrenheit cannot be maintained. The sprinkler piping system is filled with pressurized air or nitrogen. Upon the activation of a sprinkler, the air or nitrogen is purged and the loss of pressure opens a clapper valve that releases the flow of water into the system. The system then operates like a wet system. The main difference between a wet and a dry system is the location where the water is held back from activation.

Dry sprinkler systems have dedicated heated rooms containing the dry valve assembly. A compressor is provided in the room to pressurize the system with air or nitrogen and hold back a clapper valve. The clapper valve holds back the water in the sprinkler system (Figure 7.11).

Figure 7.11 Typical Arrangement within a Dry Valve Sprinkler Room. Photograph by Rodrigo Fernández. *Installed nitrogen generator*. June 10, 2013. Wikimedia Commons. Licensed under Creative Commons Attribution 3.0 (CC BY 3.0). License details: https://creativecommons.org/licenses/by/3.0/.

7.4.3 Pre-action Sprinkler System

Pre-action sprinkler systems activate under prescribed circumstances, slightly different than traditional wet or dry systems. These systems utilize a pre-action valve to withhold water in the system piping under normal conditions. Additionally, fire detection is provided within the pre-action zone either by smoke detectors, heat detectors or another detection source.

Activation may require two means of detection to occur in order for the sprinkler system to activate. This additional verification ensures that sprinklers will not operate unwantedly. This prevents water flow for the accidental activation of a sprinkler head from physical damage or a defective sprinkler head. Below are the common types of pre-action systems:

Non interlock system

A non-interlock pre-action system will activate when a sprinkler head is activated or when the secondary detection system is activated. It only needs one of the two to activate the system.

Single interlock system

A single interlock system requires the activation of the associated heat or smoke detector to be activated to allow the pre-action valve to open and allow water to flow into the pre-action zone. From this point forward, the sprinkler system operates very similarly to a standard wet sprinkler system. Water will be held back until the second step of a sprinkler reaching its temperature rating and the bulb or fusible link breaking occurs, allowing the flow of water.

The ideal situation is that the smoke detector or heat detector is activated before the sprinkler is. Therefore, the temperature rating of the heat detector is typically less than the temperature rating of the sprinkler. This allows for the ideal sequence of the heat detector activating first, allowing water to flow into the sprinkler system. This allows the flow of water into the sprinkler system and prevents any delay in water delivery.

This type of system significantly reduces the risk of unwanted sprinkler waterflow from physical damage to a sprinkler head. However, since water is entered into the system upon the activation of a smoke or heat detector, before the activation of a sprinkler, there is the risk of this water leaking through pipe connections, sprinklers and other system components. A properly maintained system can overcome these challenges by ensuring the system is tested and inspected properly and that no leaks in the system are present.

These types of systems are used in sensitive areas of the building where sprinkler water could damage expensive or critical building systems. They are also utilized when protecting irreplaceable items that hold sentimental value such as artwork, data, video, and music archives. Transformer vaults, main switchgear rooms, data centers and art galleries are common examples of where these types of systems are frequently used.

Double interlock system:

A double interlock system requires two events to occur before water is entered into the pre-action sprinkler zone:

1 Activation of the associated heat or smoke detector.
2 Activation of sprinkler within pre-action zone.

The double interlock system provides the best protection against unwanted or accidental sprinkler waterflow. It also provides the best protection against potential leaks in the sprinkler system that might be possible in a single interlock system.

These types of systems can be used in facilities with extreme cold temperatures. This prevents water from ever entering the system unless a real hazard, such as a fire, is present. This prevents ice buildup in the system and prevents the need to thaw the system and area. These systems can also be used where single interlock systems are used: transformer vaults, main switchgear rooms, data centers and art galleries. However, some jurisdictions prohibit double interlock systems due to the delay in water delivery or some may only permit them in cold, freezing facilities.

7.4.4 Deluge Sprinkler System

Deluge sprinkler systems are similar to pre-action systems since they have a valve, a deluge valve, holding back water from the sprinkler system and require activation from a smoke or heat detector to activate the system. However, unlike pre-action systems, the sprinkler system is not pressurized and all sprinkler heads within the deluge zone are open. This means that when the associated heat detector(s) or smoke detector(s) are activated, the deluge valve opens and water enters the system. Water is then discharged from all sprinkler heads within the zone, without requiring further activation of an automatic sprinkler head's glass bulb or fusible link. This delivers a great deal of water to the protected zone. This system is used in high-hazard facilities to quickly extinguish and prevent the spread of fire. Deluge systems can be used in aircraft hangars, chemical processing facilities, power plants and other high-hazard uses.

7.4.5 Foam Systems

A less common, but highly effective fire suppression system is a foam-based system. These operate similarly to a standard system; however, a foaming agent is added to the system. The expansive foam discharges over the protected area through nozzles and smothers the entire area. The foam extinguishes the fire by limiting oxygen to the fire and cools surrounding areas to prevent the spread of fire.

There are different types of foaming agents, and the concentration of the agents is engineered to be tailored for specific hazards. Low, medium and high expansion foams are chosen to alter the viscosity of density of the foam discharged to meet project needs.

The system can be activated by various detection systems. The foaming agent is stored separately in a tank and is only introduced to the water before discharge when the system is activated.

These types of systems can be installed in aircraft facilities, marine applications, buildings with significant flammable liquids and other high-hazard buildings.

7.4.6 Clean Agent Systems

Clean agent systems utilize a gaseous suppressant to extinguish a fire. These systems are typically designed to flood the protected area by releasing the gaseous suppressant and displacing the oxygen present. These systems can be activated manually, by pulling a pull station, or by automatic means by a fire detection system. In either case, due to the displacement of oxygen approach, a warning system is sounded to allow occupants to exit the space before the system is activated with the goal of displacing the oxygen with the gaseous suppressant.

The gaseous suppressants are stored in pressurized tanks near the protected zone. When the system is activated, these pressurized gases expand and flood the protected areas. Inert gases such as nitrogen or argon can be used, or carbon dioxide or halocarbon agents.

A huge advantage of these types of systems is the fact that water is not the means of extinguishment. The gaseous suppressant is able to extinguish the fire without leaving significant water damage in the area. Residue from the clean agent is minimal, if any; therefore, contents of the protected area that weren't damaged by fire are salvaged from any further damage from water.

Furthermore, these gaseous suppressants are nonconducting, so they are very suitable to protect against electrical hazards.

7.4.7 *Dry Agent Systems*

A dry agent suppression system is one that utilizes a dry-chemical powder to extinguish a fire. These systems have large tanks that store the dry chemical. When detection of a fire occurs, the dry chemical is discharged through nozzles over the protected areas.

The dry chemicals are nonconducting, so they are also very suitable for electrical hazards. The dry chemicals typically used are sodium bicarbonate and monoammonium phosphate. A dry chemical is chosen based on the specific hazards present.

However, due to the significant powder discharged when activated, the residual damage and cleanup can present challenges.

7.5 Fire Standpipe History

In the 1880s, the United States saw the onset of a construction boom of "skyscrapers." Up until this point in history, buildings in cities were predominantly low-rise structures. However, due to advancements in design and construction and new building materials, it became possible to build taller. A much lighter-weight structural system was introduced: iron and steel. This lighter-weight structure allowed the soil bearing capacity to support taller buildings opposed to the heavy masonry structures.

In the aftermath of the Civil War, there was an economic boost that supported an expansion within cities. New York and Chicago were the first cities to see these types of developments. In Chicago, the Home Insurance Building was completed in 1885, standing ten stories tall and 138 feet. Two additional floors were added in 1891, making the structure 12 stories and 180 feet. This building has been coined as the first skyscraper in the world.

In Manhattan, New York, several skyscrapers were designed and constructed in the 1890s and early 1900s. The Park Row Building was constructed as an office building located at 15 Park Row. Construction was completed in 1899. The building has 26 floors and two occupiable basement levels. There are also two cupolas, or towers, that extend up at the two front corners of the building, reaching a 31st-floor level. The building was constructed with structural steel.

Another notable, tall building constructed in this timeframe was the Flatiron Building in New York City. It was completed in 1902, standing 22 stories tall and 285 feet. This building was located on a triangular site, with the building matching this same geometry (Figure 7.12).

Vertical construction was taking off in the early 1900s and provided many benefits. It was now possible to build bigger buildings with more stories, which meant more area, resulting in more leasable floor area, and therefore higher rents. These could be built on the same sites that were limited to much smaller buildings just years before. They also shaped the skylines of major cities, extending high above the low-rise buildings that blended in together. However, with these tall structures came life safety and firefighting concerns. Fires on upper floors of buildings posed significant challenges to fire departments.

Figure 7.12 Photograph by Unknown author. *Flat Iron Building, New York, N.Y.* August 1909. Public domain. Library of Congress.

The most considerable challenge with these tall buildings was supplying firefighting water to upper levels. The use of many fire hoses, extending up several flights of stairs, was required. This approach requires a lot of equipment, manpower and delay in carrying out. Another issue was getting adequate water pressure and flow at the level of the fire. The water pressure loss, typically referred to as head loss, decreases by 0.433 pounds per square inch (PSI) per foot of elevation change. That means that water pressure at 200 feet above street level would be decreased by 86.6 PSI due to no other factor than elevation change. This left the upper stories of buildings extremely vulnerable due to a lack of available adequate water pressure and no internal water supply for firefighting efforts. These issues demanded a solution. The solution would be to provide an internal water supply dedicated to firefighting efforts.

7.6 Standpipe System Components

Standpipes are a feature of a building's fire protection system that provides a network of piping with hose connections or hoses for use in firefighting efforts. They serve as a dedicated internal water source or hydrant to provide water to a hose, which is used to spray on a fire. They typically extend vertically in buildings, often located within exit stairwell enclosures. However, in some cases, horizontal sections are run to provide hose connections to remote areas or selected areas of the building. There are several different classes and types of standpipes that change the way the systems operate and are used.

7.6.1 *Class I Standpipe System*

Class I standpipes are systems that provide a 2–1/2" hose connection for use by firefighters and trained individuals. These systems do not provide hoses; rather, they have connections that are suitable for common fire hoses. These systems are typically installed within fire-rated, exit stairwell enclosures. Most codes typically require them in each required exit stairwell. A hose connection is provided at each floor level, either at the intermediate stair level or the landing. Codes and fire department preferences will dictate the location. Class I systems are the most common type of standpipe system in fully sprinklered buildings in today's world. They provide fire departments with high-pressure and water flow to support firefighting efforts (Figure 7.13).

7.6.2 *Class II Standpipe System*

Class II standpipes are systems that provide a 1–1/2" hose station for use by building occupants or by firefighters during initial efforts. These systems provide hoses connected to the standpipe system for immediate use, without the need for additional equipment or hoses. The hoses are typically 100-foot lengths, folded neatly and stored within cabinets. These systems have minimal pressure and water flow compared to a Class I system. To use the Class II standpipe, an individual opens the

Figure 7.13 A Class I, Wet Automatic Standpipe Connection. Photograph by author.

cabinet, deploys the hose and turns on the water via a valve in the cabinet. These systems are installed so all portions of the floor are within the hose length of the standpipe. Additional standpipe cabinets are installed throughout the floor to ensure all areas are served. Class II standpipes are being phased out due to the evolving fire protection approach in buildings. Since buildings are now equipment with several fire protection systems, providing hoses for building occupants to use is no longer appropriate. It has become more of a liability than a benefit.

7.6.3 Class III Standpipe System

Class III standpipes are a combination of Class I and II standpipe systems. They provide a 1–1/2" hose station for use by building occupants and a 2–1/2" hose connection for fire department use. Therefore, they contain hoses connected to the standpipe system for immediate use, without the need for additional equipment or hoses by building occupants. The hoses are typically 100-foot lengths, folded neatly and stored within cabinets. They also include a hose connection that is suitable for common fire hoses that the fire department would connect. To use the Class III standpipe as a building occupant, an individual opens the cabinet, deploys the hose, and turns on the water via a valve in the cabinet. To use the Class III standpipe as a firefighter, the cabinet is opened and the fire hose is connected to the

2–1/2" hose connection, similar to the Class I arrangement, then the control valve is turned on. These systems are typically installed within fire-rated, exit stairwell enclosures and installed in remote areas if needed to ensure all portions of the floor are within the hose length and spray distance of the hose. These systems are usually installed in non-sprinklered buildings as governed by codes and standards.

7.7 Types of Standpipes

There are different types of standpipes that primarily differ from a wet or dry system and automatic versus manual.

7.7.1: Automatic Wet – this system has a water supply connected that is capable of delivering water to the most hydraulically demanding hose outlet. To operate the standpipe, the opening of the control valve at the desired hose outlet is the only step required to deliver water immediately. This is the most common type of standpipe system and the most reliable.

7.7.2: Automatic Dry – this type of system operates most closely to a dry sprinkler system, where pressurized air normally occupies the piping system. A dry valve holds back the water supply. Only when a drop in pressure occurs by opening a control valve will the dry valve be able to open and release water flow into the system. These systems are utilized when the standpipe is located in an unconditioned space where there is a risk of freezing. Parking structures, loading docks, terraces and roofs are areas where these types of systems may be specified.

7.7.3: Manual Dry – this type of system does not have a water supply system connected to the system. A connection from the fire department's pumper truck and/or hydrant is required to provide the water source. These systems are often utilized on construction sites when the building does not yet have a water supply connection and is not yet conditioned. While this does provide an important firefighting tool to firefighters, it does require them to connect the water source for the system to have any use. Even when a water source is provided to the system, it takes time for the system to fill up before water can actually be used from a hose outlet, causing additional delays.

7.7.4: Manual Wet – this type of system is similar to the manual dry, except that the standpipe is left filled with water sitting in the system without a permanent water source. This system would only be appropriate where the building can be maintained above 40 degrees Fahrenheit to prevent freezing. A small advantage this system has over the manual dry-type is that once a water connection is made to the system, there is not a delay in the delivery of water like the manual dry-type system.

7.8 Fire Department Connections

Fire Department Connections, or commonly referred to as FDCs, are a critical component of the standpipe system and sprinkler systems. This is the connection point at which the water supply is supplemented, in the case of an automatic style system. In the case of manual style systems, this is the connection point where water is introduced to the system altogether.

The location and other installation parameters are dictated by building and fire codes. However, these are typically located at street level, along the building's exterior wall. The number of inlets, or connections, can vary from a single to a 6-way (six inlet connection points). There is also an additional indication on the FDC that signifies what the connection serves. For example, a designation is made between sprinkler systems and standpipe systems. If the FDC does not serve the entirety of the sprinkler or standpipe system, further signage is provided to make clear and obvious the connection points for various building areas. For example, parking levels may have a dedicated FDC separate from the rest of the building. Figure 7.14 shows a typical arrangement of a fire department connection.

Figure 7.14 Typical Fire Department Connection. Photograph by author.

7.9 Shared Components in Fire Sprinkler and Standpipe Systems

Fire sprinkler systems and standpipe systems operate in a similar manner and require similar piping components. Therefore, to maximize efficiency, many components in a building's fire protection system serve both fire sprinkler and standpipe systems. The water supply, fire pump and sprinkler riser are all dual-use elements serving both systems. The fire pump delivers the required water flow and pressure to serve the hydraulic demand of the system, including the most hydraulically demanding sprinkler design area, and an allowance for a hose stream by the standpipe system. The network of piping then navigates to the exit stairs, which typically house the vertical fire protection piping in the building. This vertical rise of fire protection piping is called a combination fire sprinkler-standpipe when it supports both functions.

7.10 Maintenance of System and Testing and Inspection

Fire protection systems must be maintained in good working order to ensure that these critical systems will operate as designed. Engineered fire protection systems are only useful if they perform the function they were designed to do. Active fire protection systems require the most maintenance, as these often include mechanical and electrical components. They also include sitting water in piping and valves that can be subject to corrosion over time. For example, fire sprinkler system components such as valves, pumps, check valves and sprinkler heads all require care.

Most jurisdictions adopt a form of NFPA 25, *Standard for the Inspection, Testing, and Maintenance of Water-Based Fire Protection Systems*. This document sets the minimum inspection, testing and maintenance requirements for sprinkler systems and their components. Specialty types of sprinkler systems, including foam, mist and others, are included as well.

The requirements are broken down into three categories. The three categories and their definitions from 2026 Edition of NFPA 25 are below[4]:

3.3.24 Inspection.

A visual examination of a system or portion thereof to verify that it appears to be in operating condition and is free of physical damage.

3.3.46 Testing.

A procedure used to determine the operational status of a component or system by conducting periodic physical tasks.

3.3.27 Maintenance.

In water-based fire protection systems, work performed to keep equipment operable.

Most ongoing testing and inspection of the fire sprinkler system occurs on a quarterly basis. Components such as gauges, control valves, waterflow alarm devices and tamper devices are inspected visually, and waterflow alarm devices are functionally tested. This can include waterflow tests, fire pump tests, alarm tests, and trip tests of dry pipe, deluge, or preaction valves.

For buildings with fire pumps, there are additional inspection, testing and maintenance requirements to consider. On a weekly basis, the fire pump room and its components must be visually inspected. This includes confirming adequate heating within the room to prevent freezing temperatures. It must also be verified that valves are in a fully open position, no leaks exist and that pressure gauges are within the acceptable ranges. Fire pumps must also be tested every week under a non-flow condition called a churn test.

Property managers must understand the minimum testing and inspection requirements the project is subject to, develop a maintenance plan with a qualified company, and carry out the plan and maintain records.

It is extremely important that all components that need to be visually inspected and physically tested are accessible. While some components may be installed in designated rooms, such as fire pump rooms, dry sprinkler valve rooms, some components are simply located throughout the building in various locations. Often, floor control valve assemblies are located within exit stairwells. This is a very logical location since the combination sprinkler and standpipe riser is located here. This allows for easy access for ongoing testing or to isolate areas if necessary when conducting impairments.

7.11 Sprinkler Valve Accessibility

Zone control valve assemblies (ZCVA) that serve specific areas of the building, or specific tenant spaces, must also be located in areas that will be accessible when the tenant spaces are both in a core and shell state and when the tenant fit-out is complete. While the typical guidance on MEP equipment is to keep it high to the deck/ceiling above, the one exception is to valves and components that require frequent inspection and testing.

Sprinkler valves and the associated tamper and waterflow devices shall be installed and maintained in accessible locations. These valves are accessed often for quarterly sprinkler testing, draining of the system during impairments, and isolating a zone after the activation of a sprinkler head. Therefore, they must initially be placed in an accessible location for easy access, but also be maintained through later building alterations. Below are some arrangements that could be implemented to ensure access:

1 Sprinkler valves shall be located in accessible locations visible from the floor without removing or moving ceiling panels, light fixtures or other fixed elements and furniture. They should be located at a maximum height of 10’–0” AFF.

2 Sprinkler valves shall be located in accessible locations with access through small hatches or panels with signage that indicates the location of the sprinkler valve. The opening or hatch must be able to accommodate an 8' ladder with additional working space for a technician.
3 Sprinkler valves located above 10'–0", must be provided with a permanent means of accessing and operating the valve, such as a ladder, stairs, and/or platform conforming to OSHA regulations.

The sprinkler ZCVA must be provided with a means to drain any trapped water unable to drain to the main sprinkler drain. A low-point drain should be provided to drain this trapped water.

References

1. Wonning, Paul R. Short History of Fire Fighting: The Story of the Fire Fighter and Fire House. Bloomington, IN: Mossy Feet Books, 2019.
2. Parmelee, Henry S. U.S. Patent No. 218564: Improvement in Automatic Fire-Extinguishers, filed December 27, 1878, and issued August 12, 1879.
3. Grinnell, Frederick. U.S. Patent No. 600246: P. Grinnell Automatic Fire Extinguisher, filed April 20, 1896, and issued March 8, 1898.
4. National Fire Protection Association, NFPA 13: Standard for the Installation of Sprinkler Systems, 2013 ed. (Quincy, MA: National Fire Protection Association, 2013), 3.3.23, 3.3.25, and 3.3.47.

Chapter 8

Fire Extinguisher Fundamentals

8.1 Fire Extinguisher History

Fire extinguishers were invented in the 18th century. In 1723, a patent was issued to chemist Ambrose Godfrey for a primitive version of a fire extinguisher.[1] In the 19th century, advancements were made closer to the design of modern fire extinguishers. In 1818, George William Manby developed an improved fire extinguisher design including a copper container and potassium carbonate in compressed air.[2] Several improvements to the design were made from these early days to modern fire extinguishers. Most modifications recently are related to the extinguishing agents, with little to no changes to the fundamental workings of the device.

8.2 Fire Extinguisher Goal

Fire extinguishers are unique in that they are present in almost all buildings and homes. They are found on boats, ships and docks. They are provided on aircraft. They are located on some large machinery. They are the most prevalent fire protection system. They are also unique in that they are expected to be operated by everyday people with differing knowledge of how they operate. Most people do have the awareness and understanding that it is a good idea to have fire extinguishers provided throughout their residence. However, it usually ends there. The effectiveness of fire extinguishers are influenced by several factors. This includes the location and access to these devices. Devices must also be regularly inspected to ensure adequate charges and functioning components. Users must also be trained in how to properly operate a fire extinguisher. A false sense of security can be created if they can't be used effectively.

These same challenges exist on construction sites. Fire extinguishers are supplied on almost every construction site. However, how they are maintained plays a major role in their effectiveness on the site. Fire extinguishers are the first line of defense against a fire. They are most effective when the fire is in its infancy. When the fire grows larger, the fire extinguisher becomes a lot less effective.

The goal of a fire extinguisher is to suppress the fire either completely or to contain it until additional extinguishing means are provided. There are several different types of extinguishers based on the expected hazard present.

DOI: 10.1201/9781032720210-10

8.3 Fire Extinguisher Ratings

Fires are classified into the following categories from NFPA 10 Section 5.2:

Class	*Types of Hazard*
Class A Fire	Fire in ordinary combustible materials, such as wood, cloth, paper, rubber and many plastics
Class B Fire	Fire in flammable liquids, combustible liquids, and flammable gases
Class C Fire	Fires that involve energized electrical equipment
Class D Fire	Fires in combustible metals, such as magnesium, titanium, zirconium, sodium, lithium, and potassium
Class K Fire	Fires in cooking appliances that involve combustible cooking media (vegetable or animal oils and fats)

The types of fires are designated by letters, A, B, C, D and K. There are size ratings designated by numbers for Class A and Class B fires that represent the size of fire it can control. The larger the number, the larger the design fire it can suppress. For Class A size ratings, the number represents the water equivalency. Each number represents 1–1/4 gallons of water. A 4A rating would represent a 5-gallon water equivalency. For Class B size ratings, the number represents the area it can cover. 10B can cover 10 square feet of a Class B hazard. For Class C, Class D and Class K, the designation is given without a size rating and is rated based on weight and other parameters.

Combination Class ABC fire extinguishers are the most common to protect against the typical hazards found in buildings and homes.

For example, a 2A:10B:C fire extinguisher signifies it is a combination fire extinguisher that can protect Class A fires with the equivalency of 2–1/2 gallons of water, Class B fires with the coverage of 10 square feet, and Class C fires. This fire extinguisher contains 5 pounds of an extinguishing agent resulting in a total device weight of about 9–10lbs.

Building and fire codes dictate the sizing, location and other installation requirements of fire extinguishers. It is important that the appropriate type of fire extinguishers is provided to adequately protect against the potential hazards. Special attention must be given to projects involving restaurants with oils and fats and to manufacturing facilities with combustible metals. The wrong fire extinguishing agent can do more harm than good, so it is critical to understand what hazards are present.

8.4 Locations of Fire Extinguishers

Fire extinguishers must be provided throughout the construction site as outlined in the construction fire safety plan. NFPA and ICC codes stipulate travel distances to fire extinguishers. In most cases, a maximum travel distance of 50 feet or 75 feet is required.

Fire extinguishers should be either wall-mounted or provided on stands with adequate signage. Access should be maintained around the extinguishers. As site conditions change, it is important that clear paths and signage are maintained to ensure fire extinguishers remain readily available in emergency situations (Figure 8.1).

Fire extinguishers must also be provided at each hot work activity location. By having a fire extinguisher readily available to the fire watch, any smoldering or fire initiated by the hot work can be immediately extinguished.

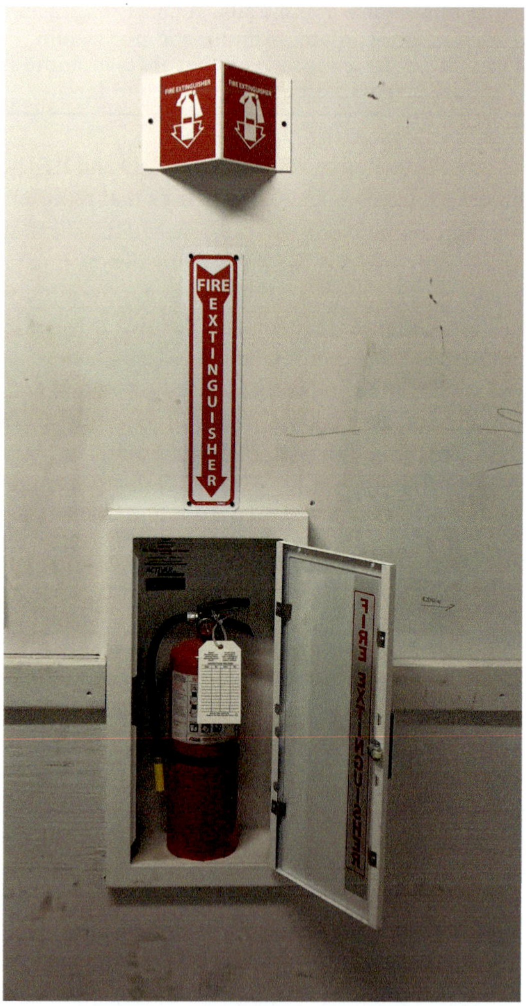

Figure 8.1 Fire extinguisher located within recessed cabinet with signage. Photograph by the author.

8.5 Maintenance of System and Testing and Inspection

Fire extinguishers installed in accordance with NFPA 10, *Standard for Portable Fire Extinguishers*, must meet the minimum inspection, testing and maintenance requirements found in this standard.

As with any system or piece of equipment, proper maintenance is required to ensure these fire extinguishers remain in good working order. NFPA 10 requires that fire extinguishers be visually inspected when placed into service and at monthly intervals thereafter. Records should be logged on the tags. Annually, a more thorough examination of the fire extinguishers is required. This includes a detailed inspection of mechanical parts, extinguishing agent, and other components on the device. An updated tag is dated and the device is put back into service. Additional testing and internal inspections are required by NFPA 10 based on the extinguisher type.

References

1. Godfrey, Ambrose. An Account of the New Method of Extinguishing Fires by Explosion and Suffocation; Introduced by Ambrose Godfrey, c. 1743.
2. Manby, George William. Improvement in Fire Extinguishers, British Patent No. 3411, filed 1813.

Chapter 9

Smoke Control System Fundamentals

9.1 The Risks Associated with Smoke

When most people envision what it might look like to experience a fire or they have ideas from how fires are depicted in movies, it doesn't always correlate to real-life fire development. A fiery, fierce flaming fire could occur and could certainly be dangerous. While the dangers of the heat and flames do pose a major threat to those nearby the source of the fire, it is the development of smoke and other toxic byproducts of combustion that pose the greatest threat to areas much greater than the point of origin. Smoke inhalation is the leading cause of death and injury for victims of fire. It is estimated that 60% to 80% of deaths at a fire scene are directly attributed to smoke inhalation.[1]

9.2 MGM Grand Hotel and Casino – Las Vegas, NV

A tragic example of the dangers of smoke spread and the toxicity of the smoke from a fire can be demonstrated with the MGM Grand fire that occurred on November 21, 1980. The building was 26 stories, consisting of casino gaming areas, restaurants, convention areas on the lower levels, and hotel rooms throughout the tower. A fire started in the early morning in a first-floor restaurant known as "The Deli." An employee discovered the fire around 7 a.m. The fire spread to adjacent spaces in the building on the first floor, including lobby areas and into the casino floor, spreading quickly.[2] At the time of the fire, it was estimated that there were approximately 5,000 people in the building, many in the 2,000 hotel rooms in the building.

The fire that spread across the first floor of the building created a significant amount of smoke. Smoke spread vertically throughout the building through exit stairwells, elevator hoistways, unprotected openings and other mechanical chases and shafts. Deficiencies were discovered in exit stairwells where there were unsealed areas and a lack of a two-hour fire-rated enclosure. This led to the fast deterioration of conditions within the exit stairs. The smoke infiltration throughout the building created extremely toxic conditions that left occupants starving for fresh air. Many occupants tried different exiting paths, only to be unsuccessful due to the stairs being filled with smoke with little to no visibility. Many occupants remained

DOI: 10.1201/9781032720210-11

Figure 9.1 Photograph by Clark County Government. *MGM Fire Pic 1 (photograph of the MGM Grand Hotel fire, November 1980), Wikimedia Commons, public domain.*

trapped in their hotel rooms, or many fled to the roof level, where there was breathable air.

The fast-developing fire prompted a large response from a number of agencies. This included local and surrounding cities' fire departments and rescue teams. It even included the United States Air Force to carry out rescue efforts to pull hundreds of trapped occupants from the roof. A total of 554 firefighters responded to the fire at various times during the incident.

A total of 85 people died as a result of this fire incident, including hotel guests and employees. Toxicology analysis of blood samples of victims showed that the majority of victims died of asphyxiation secondary to carbon monoxide inhalation. In fact, all 74 casualties in the high-rise tower were asphyxiation secondary to carbon monoxide inhalation. This highlights the danger that smoke spread can have on occupants in a building (Figure 9.1).

9.3 Smoke Control System Introduction

Smoke control systems are used to manage the spread of smoke within buildings. The management of smoke from a potential fire is critical to protect occupants in the building. While fires can be widespread, often they are contained within a single compartment of a building, such as a single room, floor, or area of a building. If smoke is not properly managed, the impact of the smoke can be far more significant than the fire itself. Therefore, smoke control systems are engineered into buildings to minimize the risks of rapid smoke spread.

The management of smoke can utilize many different systems, infrastructure and other design considerations. The following sections outline some common systems seen in buildings today.

9.4 Atrium Smoke Control Systems

Large atriums and other vertical floor openings connecting multiple floors present a major smoke spread risk. A single fire within an atrium can produce smoke that can quickly disperse to other connecting floors. Buildings typically rely on passive fire protection systems to compartmentalize the building. This includes elements such as rated floor and ceiling assemblies and rated door openings. Penetrations are firestopped to maintain fire and smoke separation between each floor level or sometimes even between different areas of a single floor. However, when vertical openings or atriums are provided in buildings, this protection cannot be counted on. Therefore, an atrium smoke control system is provided to overcome these associated risks.

When a fire occurs, smoke is produced as a byproduct of combustion. Within a large atrium space, this smoke plume rises to the highest ceiling plane. The smoke gathers and spreads horizontally across the ceiling, developing a smoke layer. This smoke layer depth continues to grow as smoke continues to be produced from the fire. It is critical that the smoke layer depth does not grow large enough that it would spread to the floor levels below and create untenable conditions for occupants.

Therefore, means of removing the smoke at the ceiling level are essential to prevent overaccumulation of smoke. Engineered systems can include a natural ventilation system where a large vent is installed at the ceiling to allow hot smoke and gases to escape. The vent is designed to open upon detection of a fire. Factors like wind, ambient indoor and outdoor air temperatures and indoor HVAC systems all impact the effectiveness of a natural ventilation system.

Atrium smoke exhaust systems in the United States primarily use mechanical smoke exhaust fans to remove smoke from large spaces. Large exhaust fans are sized and located based on engineering analysis of the space. Often, computer-based fire modeling tools are used to assist with the engineering of these systems. The use of computational fluid dynamics tools helps anticipate the temperatures expected and predict smoke spread from different fire scenarios, often utilizing a worst-case scenario, to design the system around.

One or more exhaust fans are located at the highest ceiling plane, or along the top of the wall just below the ceiling. These fans exhaust the smoke from inside the atrium space to the exterior when activated. Make-up air must be provided to allow for the exhaust fans to operate effectively since a significant amount of air is being exhausted by the large fans. Make-up air can be provided from mechanical air supplies or can be from natural means through automatically opening doors and windows.

The activation of the atrium smoke control exhaust system is by an automatic fire detection system. Projected beam type smoke detectors are commonly used in large open areas, like atriums, to detect smoke from fires. Beam detectors can be calibrated to a specific obscuration percentage to detect the fire within the required performance criteria. Other detection methods, such as open area smoke imaging detection (OSID) or other detection methods may be chosen by the engineer.

Other components of an atrium exhaust system may include smoke and fire curtains that automatically deploy upon the detection of smoke. These systems can provide fire and smoke-rated separation between the atrium and the remainder of the building. Other passive fire protection systems, such as rated wall assemblies and doors, can provide separation between the atrium and the rest of the building as well.

The overall atrium smoke control system includes many components such as fans, makeup air, automatically opening doors and windows, smoke detection systems, deployable fire and smoke curtains and other passive fire protection systems. Therefore, when maintaining these types of systems, it is important that all elements of the system are regularly tested and inspected.

9.5 Smoke Proof Stair Enclosures

In multistory buildings, aside from the level of discharge, exit stairs are the only means of getting occupants from the above or below-grade floor level to the level of discharge and to the exterior of the building. Therefore, the exit stairs are safe havens that need to be reliable in a fire scenario. Exit stairs usually have a 1- or 2-hour rating based on the number of floor levels they extend. Doors into the exit stairs are fire-rated with automatic closers.

In an existing condition, exit stairwell doors will be constantly opened and closed to allow occupants on each floor level to enter the stairwell. This constant opening and closing allows an opportunity for smoke to enter the stairwell and traverse through the building like a chimney. This would negate the benefits of the other passive fire protection systems in place; therefore, a system must be introduced to protect against this challenge.

Although there are a few options when it comes to providing smoke-proof enclosures for exit stairs, the most common method is the stair pressurization method. Other methods are possible, such as providing exterior natural ventilated balconies; however, these are less common as they are much harder to incorporate into a functional building layout.

Stair pressurization systems work on the basis of introducing a positive pressure differential to prevent smoke from entering the exit stair. The interior of the stair is designed to have a slightly higher pressure as compared to the adjacent space, directly outside the stairways within the building. A positive pressure within the stairwell is created by stair pressurization fans. The fans supply air into the stairwell enclosure. The two main supply air methods are a single injection point system and multiple injection point systems (Figure 9.2).

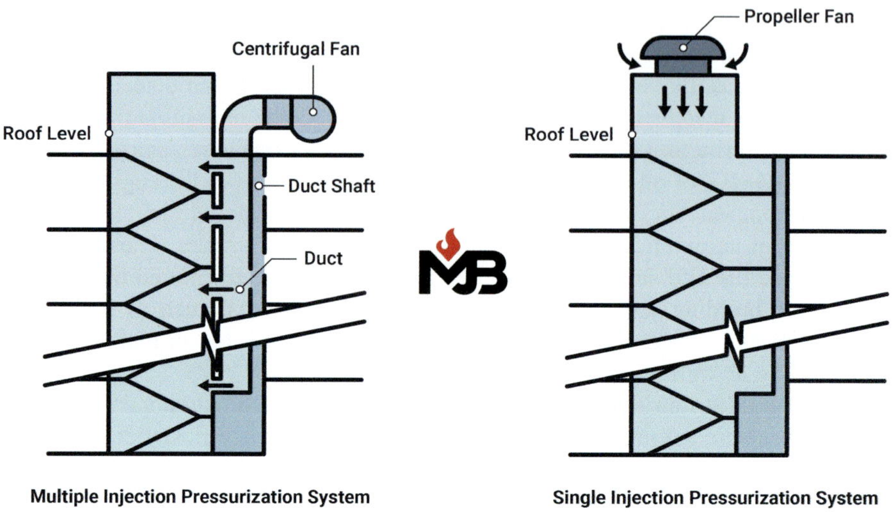

Figure 9.2 Multiple- and Single-Injection Pressurization Systems. Image illustration by St. Germain Design Co.

A single injection point is where the air is supplied into only one location within the stairwell. This arrangement has its challenges since there can be a loss of pressure if a door near the injection point is opened for too long. Other challenges, such as pressure gradients within the stairwell enclosure, may become a problem as well. In order to achieve the necessary minimum pressure differential within the stairwell, there may be pressures exceeding the maximum near the injection point to achieve the minimum pressure at the other end of the stair shaft. Therefore, a single injection method may only be suitable for stairs in low to mid-rise buildings.

More commonly, a multiple injection method is used to more evenly distribute the air supply within the stairwell enclosure. Multiple fans can be provided, spanning the stairwell, or a single large fan can be used with a ducted riser with outlets in several locations within the stairwell enclosure.

Governing codes set minimum and maximum pressure differentials across closed doors in stair pressurization systems. Typical minimum pressure differential (from within the stairway shaft, measured across the stairway door, to the exterior of the stairway) of 0.10 inches of water and a maximum pressure of 0.35 inches of water are required.

When a positive pressure is introduced to the stairwell enclosure, the exit stairwell doors are subject to an additional force. Since most doors for egress purposes swing into the exit stairs, this means the push force to open the doors may be increased when the stair pressurization system is operational. Therefore, door opening forces are measured both when the system is off and when it is operational to ensure compliance with building and accessibility codes. Conversely,

Figure 9.3 Stairwell pressurization inlet grill. Photograph by author.

exit discharge doors typically open outward (swing out of the exit enclosure). These exit discharge doors have the opposite issue, where the pressure assists in the door swinging open, sometimes making it difficult for the door to return to a closed position. The door must fully close and latch to ensure the door does not become a leak in the system that jeopardizes the pressure. It is important to ensure doors into the stairwell enclosure are fine-tuned to ensure they properly self-close and latch to maintain the barrier of the system that will allow the system to work as designed (Figure 9.3).

The building's stair pressurization system activates upon the detection of a fire alarm. Smoke detectors, heat detectors, manual pull stations and sprinkler water flow alarms will all initiate the stair pressurization system.

The overall stair pressurization system is engineered specifically for each stair's configuration and geometry. Environmental factors also influence the design of the system.

Stair Design Considerations:

1 Volume of stairwell enclosure
2 Number of floors that the stairwell serves
3 Quantity and area of doors into the stairwell
4 Expected leakage rate based on expected gaps at doors

Building and Environment Considerations:

1 Stack effect (temperature differential between conditioned interior space and outside temperatures)
2 Climate
3 Wind factor
4 HVAC system operation
5 Design fire Impacts

9.6 Elevator Hoistway Smoke Control Systems

Elevator shafts are among the largest vertical shafts in a building. Large buildings often have several elevators serving them. These vertical shafts often connect every floor of the building. Therefore, careful consideration is given to how these shafts are protected against fire and the spread of smoke.

Similar to other vertical openings like stair shafts, elevator shafts typically have a 1- or 2-hour fire rating enclosing the shaft. Openings into the shaft have fire ratings to maintain the integrity of the rating.

Smoke control systems for elevators are typically provided by hoistway pressurization systems or by enclosed elevator lobbies.

9.7 Elevator Hoistway Pressurization Systems

Elevator hoistway pressurization systems operate very similarly to stair pressurization systems. Supply fans provide air into the elevator hoistway in a single injection or multiple injection method to create a pressure differential so that the hoistway is a slightly higher pressure than the rest of the building to prevent smoke from entering.

The elevator hoistway pressurization system is engineered following a similar approach to the stair pressurization system. Supply fan sizes and locations are engineered to ensure the target pressures can be achieved throughout the hoistway.

An advantage of the elevator hoistway pressurization method, as compared to the enclosed lobby approach, is the benefit of not having a dedicated lobby outside of the elevator hoistway on each floor level. Therefore, the pressurization method can be more efficient from a building area consumption standpoint.

9.8 Elevator Lobbies

Another method of protecting against the spread of smoke at elevator hoistways is by installing enclosed elevator lobbies at each floor level surrounding the elevator hoistway door(s). The enclosed lobby enclosure is constructed as a fire-rated and smoke-rated assembly with doors into the lobby meeting similar requirements.

This method of elevator hoistway protection does require dedicating an area of the building on each floor to serve as the elevator lobby. Additional infrastructure, including smoke and fire-related doors and partitions, must be provided. However, this prevents the need to install an elevator pressurization system, including additional engineering, costly supply fans, ductwork, associated dampers, modules, relays, smoke control panels and other equipment. It also saves much effort on the maintenance of the pressurized system. The elevator lobby approach is the prescriptive approach to the elevator smoke control requirement.

9.9 FPPM Considerations

It is important to understand the factors that affect the smoke control system. It would be easy to think that if the modification does not directly impact the stairwell, no adverse effect would be expected. However, due to the impact of several external factors on the smoke control system's successful performance, all alterations need to be analyzed for their impact on the system. Therefore, if the building is undergoing renovations that include altering the layout of a floor with the addition or removal of partitions, modification of any building HVAC system, the impact on the smoke control system must be considered. The alteration of the building's HVAC system may impact the pressure differential that was originally designed. The smoke control system may be designed to shut off the building's HVAC system upon detection of a fire alarm and activation of the smoke control system. Therefore, new HVAC systems must follow a similar sequence and be provided with the necessary relay modules to allow this function to occur. The addition or removal of a partition may restrict or increase air flow, also impacting the smoke control's performance.

It is essential to always consider the impact on the smoke control system to ensure the successful operation. The engineer of record or subject matter expert should be consulted to ensure the intended operation of the system is not jeopardized by the proposed alterations.

The smoke control system depends on doors into the stairwell enclosure to be closed and latched at all times, except when occupants are entering. When occupants enter through the exit stair, the door must be able to automatically close fully and latch. If doors into pressurized stairwells are propped open or the closer on the door is defective, these large unaccounted-for openings in the engineered pressurized stair will compromise the system. The propped doors also jeopardize the fire rating integrity of the stair as well. The FPPM's careful management and monitoring of smoke control systems will ensure these systems operate to provide occupants with the increased safety that is critical in emergencies.

References

1. Anseeuw, Kurt, Nicolas Delvau, Guillermo Burillo-Putze, Fabio De Iaco, Götz Geldner, Peter Holmström, Yves Lambert, and Marc Sabbe. "Cyanide Poisoning by Fire Smoke Inhalation: A European Expert Consensus." *European Journal of Emergency Medicine* 20, no. 1 (2013): 2–9. https://doi.org/10.1097/MEJ.0b013e328357170b
2. Best, Richard. *The MGM Grand Hotel Fire: The Impact of Fire Protection on the MGM Grand Hotel Fire*. Prepared by the National Fire Protection Association (NFPA), with consultation by David P. Demers, P.E. Revised January 15, 1982. https://www.firefighternation.com/wp-content/uploads/2016/11/lasvegasmgmgrand.pdf.

Chapter 10

Passive Fire Protection Features

10.1 Passive Fire Protection Introduction and History

Passive fire protection systems are features designed and installed in buildings to intentionally protect occupants and the building from the effects of fire and smoke. This is mainly achieved by compartmentalization. Passive systems create fire-rated or smoke-rated separation to limit spread throughout different areas and levels of the building to keep a fire contained within a single area or zone. They differ from active fire protection systems by being components that do not operate or activate in the event of a fire, but rather are inherent to the building. For example, a fire sprinkler system will react to a fire by providing water to the fire location and notifying the fire alarm system; therefore, this would be considered an active fire protection system. Whereas a fire-rated, floor-ceiling assembly creates separation between floor levels, which limits fire and smoke spread. This passive system provides many benefits to the life safety level in the building.

In the 19th Century, many urban areas in the United States, Canada and beyond were seeing major conflagrations that wiped out significant portions of cities. What would start as a small fire would quickly grow into a massive blaze. The fire would spread throughout the original building, then catch adjacent buildings on fire and spread throughout the districts of cities.

The Great Boston Fire of 1872 exemplified this issue. A fire started in a commercial warehouse basement in the downtown area of the city on November 9, 1872. The fire spread to 65 acres throughout Boston's downtown and destroyed 776 buildings.[1] The fire jumped from building to building, making it extremely challenging to combat. The lack of sufficient separation between buildings, the lack of fire-rated elements, the lack of firefighting equipment and the lack of adequate water supply created a recipe for a disaster. It caused the equivalent of $1.7 billion in damages, making it one of the most devastating and costly fires in American History. There is a great book detailing the events of this fire by Stephanie Schorow titled "The Great Boston Fire: The Inferno That Nearly Incinerated the City."

It was becoming abundantly clear that considerations were needed to limit the spread of fire to other areas in the building and to nearby buildings. Many cities reacted individually after these horrific fire events by imposing construction and

DOI: 10.1201/9781032720210-12

building material regulations. However, it wasn't until the early 20th century that the first standardized building code, the Uniform Building Code (UBC), was published in 1927, focused on separation between occupancies.

As building and fire codes progressed in the 20th century, among other fire protection considerations, fire-rated assemblies and the compartmentalization strategy became among the most important considerations. Several building features and terminology were established to detail fire and smoke-rated separation between floor levels, building areas, stairwells, and other areas.

10.2 Fire and Smoke Rated Construction

10.2.1 Fire Walls

Fire walls are probably the most recognized passive protection system for people in and out of the construction industry. A fire wall is a fire-rated wall structure, typically extending from the foundation of a building vertically through the roof. A fire wall is one of the most robust passive fire protection systems, both in the way it is designed and its structural integrity. A fire wall must be constructed so that it is structurally independent from both sides of the building, so if one side of the building collapses or is structurally compromised, the other side of the building and fire wall is still intact. Fire walls must also extend horizontally to reach exterior walls of the building, to ensure the fire cannot jump around the wall at any point vertically or horizontally.

Fire walls can be constructed of various materials to achieve the desired fire rating of the wall. They can include masonry components such as concrete masonry units (CMU), brick or concrete. They can also be constructed from metal or wood framing with multiple layers of gypsum board covering to achieve a specific rating. The fire rating of fire walls can be between 2 to 4 hours.

Most codes recognize areas separated by fire walls to be considered separate buildings due to the major benefit of compartmentalization with these features. Therefore, large buildings can be separated by fire walls to increase the area allowed by building codes. Fire walls can also be used to limit the fire exposure of a building to nearby buildings and structures. This significantly reduces the exposure risk of a building being ignited by nearby buildings.

10.2.2 Fire Barriers and Fire Partitions

Fire barriers are also a fire-rated wall assembly element; however, there is more leniency on how these can be constructed when compared to a fire wall. While they can extend up from a foundation, they can also start from a floor/ceiling assembly on any given floor and then extend vertically in the building. They can extend horizontally to exterior walls or can create a self-contained enclosure, such as a shaft or stairwell.

Fire barriers are typically used for enclosing exit stairs in fire-rated construction. They can be used for enclosing vertical run mechanical, electrical and plumbing equipment and piping in shafts. They can also create separation for rooms and spaces in buildings for lab control areas, high hazard rooms, and other rooms that require ratings, such as fire pump rooms, fire command centers and similar.

Fire barriers can be constructed of masonry materials or with gypsum wallboard and standard metal or wood framing. The fire rating of fire barriers can be between 1 to 4 hours.

Fire partitions are similar to fire barriers; however, the fire rating is typically less, in the range of ½-hour to 1-hour rating. Fire partitions are often used in corridors and interior walls to create the desired separation.

Fire-rated glass can also be utilized as part of fire-rated assemblies. Performance-based designs, such as sprinkler-protected glass, allow for the use of closely spaced sprinklers along a glass partition. The sprinklers provide a quick response and provide water along the glass to prewet and cool down the glass surface to achieve an equivalent fire rating.

10.2.3 Horizontal Fire-Rated Assemblies

In large, tall buildings, there is the benefit of having some built-in compartmentalization by way of multiple floor levels. When these floor and ceiling assemblies are rated, this provides a major benefit in the effort of separating floor levels should a fire event occur on a single floor. Horizontal fire-rated assemblies are those fire-rated assemblies that extend horizontally through a floor and ceiling assembly throughout a building to create separation above and below.

Some of the major triggers to the required rating of a floor/ceiling assembly is the construction type of the building. The construction type of a building is determined by the size of the building (both area, height and number of stories) and the use groups contained in the building. Type I Construction, as well as IIA, IIIA and VA, all require either a 1- or 2-hour floor construction rating.

There may be other reasons to create horizontal separation in a building, such as separation between occupancies, laboratories, residential dwelling units and other higher hazard areas.

Horizontal fire-rated assemblies can be constructed out of concrete, masonry products, or gypsum products. A common example of a fire-rated horizontal assembly would be a composite concrete deck. This could consist of a corrugated metal deck, rebar or wire mesh as a reinforcer, and concrete poured into the fluted pan. When the concrete cures, it creates a strong, bonded assembly. This serves as a structural floor component but is also used to achieve the required ratings for floor assemblies.

10.3 Fireproofing

The primary structure of a building, including columns, beams, secondary members and bearing walls, can either be protected or unprotected. The construction type of a building is the primary influence on this: Type I Construction, as well as IIA, IIIA and VA, all require a 1-to-3-hour fire rating for the primary structural frame. When structural steel members are used for the primary structural frame, they can either be left unprotected or protected by an application of a fire-retardant system.

Steel is extremely useful for the construction of large buildings since it is relatively light and provides much strength, as compared to concrete, brick or other heavy building materials. However, structural steel's strength can diminish if exposed to increased temperatures caused by fire. According to the American Institute of Steel Construction (AISC), steel loses 50% of its capacity at 1100 degrees Fahrenheit and loses nearly all its strength at 2700 degrees Fahrenheit.[2] Therefore, it is important to protect steel from the effects of a fire and insulate it from the potential high temperatures to maintain the structural integrity of the building during a fire. The most common way of encasing structural steel to provide a fire rating is to apply a spray-applied fireproofing product or intumescent coating.

Spray-applied fireproofing is composed of various substrates, including fiberglass fibers, sand, cement and gypsum binders. It is pre-mixed and applied to various structural elements by a sprayer. The spray-applied method allows the coating to conform to the geometry of various elements such as columns, wide-flange beams, bar joists, and other elements (Figure 10.1).

Figure 10.1 Spray-applied fireproofing on structural steel. Photograph by the author.

10.4 Firestopping

Firestopping systems maintain the rating of the penetrated assembly or joint formed between walls, floors and ceilings. Although several fire-rated elements are constructed into buildings to develop separation, there must be a way to allow building system infrastructure to pass through these assemblies. The firestopping detail must meet the rating of the wall. A 2-hour firestopping detail is needed for a 2-hour fire-rated assembly, 1-hour firestopping detail is needed for a 1-hour rated assembly and so on (Figure 10.2).

For example, in an exit stair enclosure, there may be a combination sprinkler and standpipe riser. The stairwell could be constructed as a 2-hour fire-rated enclosure. At each floor, a sprinkler pipe must be fed off the riser to provide sprinkler water for the floor. This means you have a pipe penetrating a rated wall assembly at each floor. In order to maintain the continuity of the fire barrier and fire rating, an approved firestopping assembly must be provided at all penetrated items. There are several listed systems based on the penetrating pipe material, wall assembly details, and size of the annular space. A common detail could include filling the annular space with mineral wool to a specified compression amount, and covering the product with an approved firestopping sealant.

Figure 10.2 A pipe penetration through a fire-rated assembly with firestopping system. Photograph by the author.

Joint systems are another type of firestopping system at the intersection of the top or bottom of a wall and a floor/ceiling system or at two dissimilar assemblies to maintain a specific rating. Another common firestopping area is the edge of the slab, between the exterior wall or curtain wall and the floor system. This gap is consistent on each floor; therefore, it is important to properly seal this to prevent an easy pathway for smoke and fire to quickly spread up the building. A listed assembly must be chosen to seal this gap that maintains the rating of the floor/ceiling assembly.

A firestopping detail is typically chosen from a listed and approved detail that was tested to applicable testing standards. These details specify several factors, including the assembly being penetrated, such as a wall or floor, and specific construction characteristics, such as metal stud framing, 16" on center, with 5/8" Type X Gypsum. The penetrated pipe or item is defined, such as a schedule 40 steel pipe. Then the firestopping detail elements are also defined to specify the products to be used. A fill material of mineral wool or similar can be specified, including the compression percentage. A sealant product is listed and so is its thickness and coverage.

For those unique arrangements, where a listed detail cannot be perfectly applied to the situation, an engineering judgment (EJ) may be appropriate. An EJ can be developed by a qualified party to essentially create a compliance alternative that addresses the specific nature of the situation.

10.5 Fire and Smoke Rated Doors

Fire and smoke-rated doors maintain the desired rating of a wall assembly while allowing for the passage of egress travel through the wall. The rating of the door must be consistent with the rating of the wall assembly or in accordance with applicable building codes, which sometimes allow a slightly lower rating in the door. The doors must be self-closing and positive latching. This ensures the doors automatically return to a closed, latched position after traversing through them.

Fire and smoke-rated doors are required to have an affixed label that displays the rating, manufacturer, test standards, and if there is an associated smoke rating associated with the door.

10.6 Fire and Smoke Dampers

Another common element that penetrates through rated assemblies is ductwork. Ductwork is run throughout buildings to provide ventilation, heating and cooling, and exhaust. When ductwork travels through either a floor running vertically or through a fire-rated wall horizontally, considerations must be made to maintain the integrity of the rated assembly. Otherwise, the ductwork could provide a route for fire and smoke to spread in a fire situation.

There are three main types of dampers:

Fire dampers – For use in fire-rated assemblies, where it is not necessary to protect against smoke spread. These typically contain a fusible link that activates when exposed to heat and shuts a damper to restrict the passage of hot gases and fire.

Smoke dampers for use in smoke-only rated assemblies – These utilize a duct smoke detector to detect smoke and activate a motorized damper.

Combination fire and smoke damper – These protect against the risks of both fire and smoke using a combination of technology.

References

1. Mass Moments. "Great Fire Devastates Boston," November 9, 1872, https://www.massmoments.org/moment-details/great-fire-devastates-boston.html.
2. American Institute of Steel Construction. "11.2. Steel Exposed to Fire." *Steel Solutions Center*. Accessed February 6, 2025. https://www.aisc.org/steel-solutions-center/engineering-faqs/11.2.-steel-exposed-to-fire/#9370.

Chapter 11

Means of Egress

11.1 Means of Egress Introduction

A building's means of egress system plays a major role in providing safe evacuation routes during fire emergencies. These systems are carefully designed to ensure there is adequate exit capacity with a sufficient number of exits. Exit access and signage are examined to ensure occupants are provided with clear, safe routes to building exits. During a fire, the fire protection systems provide additional time for occupants to evacuate the building, but the means of egress systems provide the actual components that allow safe passage to a safe haven. This is referred to in the fire protection engineering world as the available safe egress time (ASET) and the required safe egress time (RSET). In performance-based design, these factors are closely studied to ensure that means of egress components are sufficient based on the actual loading of the building and geometry of the floor plans. It is important that all involved in real estate, even if they are not designing means of egress, have a basic understanding of means of egress systems. Events, modifications or improvements to the building must not diminish these critical systems.

Means of egress are comprised of three components:

1. Exit Access
2. Exit
3. Exit Discharge

The design of the means of egress for a building, including these three components, is primarily driven by applicable code requirements or, in some cases, a performance-based design.

11.2 Occupant Load Determination

The first step in designing a building's means of egress system is to first determine the occupant load of the building, or how many people are expected based on the use of the space. The most common way of doing this is prescriptively by applying

DOI: 10.1201/9781032720210-13

an occupant load factor, typically expressed in an occupant per square footage manner. Egress analysis is completed for specific layouts in buildings. Each space and room is analyzed, and an occupant load factor is applied that most closely resembles the actual use of the space. For example, in an office building, various occupant load factors, such as assembly (for meeting areas or conference rooms) or storage or mechanical rooms, are applied in addition to the business (office) occupant loads.

Chapter 10 of the International Building Code, as adopted with amendments in many jurisdictions, provides occupant load factors for several common building uses.

11.3 Required Number of Means of Egress and Sizing

Once the total number of occupants is calculated, the next step of the design process can happen. This step determines the required number of exits and the sizing of these components. Most building codes and life safety codes will set thresholds for the required number of exits based on the proposed occupant load of the space or story.

- 1 Exit – a single exit space or building is sometimes permitted by applicable codes when the occupant load is beneath a set amount, often 49 occupants, and other features are met, such as meeting exit travel distancces and fire protection systems.
- 2 Exits – it is fair to assume in most cases that at least two exits are required from every space and every story, unless an exception is met that specifically allows a single exit. Two remotely spaced exits can typically support occupant loads up to 500.
- 3 Exits – occupant loads from 501-1000 typically require three remotely located exits.
- 4 Exits – buildings with occupant loads greater than 1,001 typically require four exits.

The requirements above for the required number of exits set the number of remotely located exits, but do not indicate the sizing of the exits. The sizing of exit components is directly correlated to the calculated occupant load. When following a prescriptive compliance method, most codes provide a minimum width of egress components per occupant for both stairs and other corridors, doors, and other components.

11.4 Exit Access

Once the number and sizing of exits are known, the next step is to locate them within the building in a manner that maintains the highest level of life safety. Factors such as travel distances, egress through adjoining rooms and spaces, tenant spaces and signage are considered. These parameters are heavily influenced by the applicable codes.

Exit access travel distances vary based on the occupancy classification of the building. High-hazard buildings typically set lower maximum travel distances compared to moderate-hazard buildings. Exits must be centrally located so that access is not required through other tenant spaces, locked rooms or spaces.

Adequate signage and egress lighting are important to ensure occupants can navigate to the provided exits. In fire scenarios, occupants tend to try to utilize the same exit/entrance they used when entering the space. However, it is important that secondary means of egress are properly lit and marked to provide additional exiting options for occupants.

11.5 Exit

The exit component itself is most commonly an exit door directly to the exterior, a rated vestibule or enclosed exit stairs. This component of the means of egress system allows the safe, protected egress from the building to the exterior. These components are sized and protected in accordance with applicable codes.

11.6 Exit Discharge

The exit discharge is the point at which the exit meets the exterior of the building. It may seem like this is the end of the means of egress system, but there are more elements that are necessary to ensure safe evacuation. The exit discharge spans from the exterior exit door to a public way. In some cases, this could be just a few steps to a sidewalk, or it could be a more significant walk through alleys, exterior steps or ramps. In either case, the entirety of the exit discharge to the public way must be adequately lit, with stairs, ramps and other components complying with applicable codes. Walking surfaces must be maintained to overcome weather conditions that could create treacherous paths.

11.7 Proper Maintenance of Means of Egress Systems

Ensuring the means of egress systems in buildings are always maintained is necessary so they are ready 24/7 to support the safe evacuation of occupants in an emergency. Locked doors, poor lighting or signage and storage within exit ways are all common issues that must be overcome. It is essential that all aisles, walkways, stairwells, access ladders, corridors, and exterior egress routes are maintained throughout construction.

The storage of construction materials or general building contents must be strictly prohibited within exits and exit pathways. The accumulation of materials is problematic in two ways:

1 It adds a fire load to a component that must be maintained as a safe haven
2 It obstructs and constrains the available clear width of exiting.

There may be enticing areas beneath stairs or within exit corridors that look like a convenient spot to store materials; however, it must be prevented to ensure the building's life safety system is not compromised.

11.8 New Construction Considerations

In new construction, you have the benefit of not having any occupants within the building other than construction workers and associated personnel to support construction operations. However, there are still many occupants found on site on any given day. On large projects, hundreds of workers, tens of supervisors, several inspectors and engineers can be found on a jobsite. Therefore, careful consideration must be made to ensure all construction personnel are provided with adequate means of egress.

In ground-up, new construction, there are many challenges to providing and maintaining means of egress systems. In multi-story buildings, often, permanent stairs lag as the building goes vertical. Temporary exits are sometimes provided by exterior stair towers or a combination of interior temporary stairs or ladders. The exits may be constantly changing as site conditions evolve. The path you knew yesterday may be blocked today without awareness by occupants on-site. These inconsistencies and changing exiting paths are a recipe for disaster if fast, safe egress is needed in an emergency.

Although many reasons led to the devastating 2023 "SouthPark" construction fire in Charlotte, North Carolina, the lack of means of egress was a major contributing factor that contributed to the two workers' deaths. The upper levels of the building at the time of the fire were only provided with a single means of egress, which was over 400 feet away from the location of the trapped workers. This creates extremely unsafe conditions for those working on the upper levels of the building. Although a secondary temporary means of egress was provided just weeks before, it was not there when it was desperately needed. It is presumed that the temporary egress stairs were removed to likely allow construction progress to continue, where this structure may have blocked the exterior façade and ground areas. However, without a plan in place, the project continued to operate in this vulnerable state before the permanent second exit stair became ready for use. The proper coordination of the means of egress through each phase could have saved the lives of those two workers.

This dilemma is mostly a logistical and means and methods challenge. The exit stairs in their permanent condition are assumed to be code-compliant and sufficient for the building as designed by the design team. However, until that permanent condition can be achieved, you are operating at a lesser level of safety.

As part of the construction fire safety plan, the means of egress must be addressed during the construction phase. A plan must be in place to ensure means of egress are adequately provided within the building at all phases of the building. Therefore, several arrangements must be considered. Excavation and foundation work may need a different solution than when the building's structure extends vertically. Priority must be given to construct and maintain exit stairs.

The construction means of egress approach must be reviewed early in the project to ensure it is integrated into the project from both a financial and a schedule perspective. It is highly recommended that permanent stairs or multiple, be accelerated during construction to provide construction personnel with sufficient exiting. Additional protection can be provided by enclosing the stairs with gypsum or other appropriate coverings and providing a door. NFPA 241 also has provisions to provide at least one-half of the required means of egress during construction with these enclosed construction stairs. So, if the building needs two exits in its permanent condition, one enclosed stair should be provided during construction. Or if three exits are required in its permanent condition, two enclosed stairs would be required during construction and so on.

This obviously has a major impact on construction sequencing since the stairs, stair enclosures and stair doors are being installed much sooner than they otherwise would. However, if a thoughtful approach is determined early in the project, many of the cost and schedule impacts can be overcome.

Providing enclosed exit stairs during construction provides the most reliable means of egress system for occupants of the building. These stairs provide a dependable and consistent exiting strategy. Exit locations and pathways are established early in the project and maintained throughout. Construction personnel learn these arrangements that are used for months instead of trying to adapt to constantly evolving methods.

Exits at the ground level must also be prioritized. This includes access to an exterior door within the building, a properly maintained swinging door, and an even walking surface to a public way free of obstructions. Two or more exits must be accelerated to allow for the safe exiting of the building. Usually, exterior site work, including final grading, happens later in the project. Therefore, it is not uncommon for exterior doors to be met with a significant drop or other obstacles during construction. Exits should be selected based on their remoteness from each other and on their access points within the building. Doors intended as exits must be clearly marked with signage and provided with adequate lighting. Equally important, doors that could be mistaken as exits that are not intended to be exits must be marked with "not an exit" signage to prevent people from using them and being faced with obstructed paths or dangerous conditions.

11.9 Existing Building Considerations

If working in buildings that are occupied during construction, maintaining the means of egress is even more important. Occupants of the building depend on the existing means of egress as a dependable escape route in case of an emergency. If egress routes are obstructed by construction materials or equipment, occupants may not be able to utilize these routes in an emergency.

A common challenge is preventing contractors from utilizing back-of-house corridors and stairwells as storage areas. This is especially challenging in smaller retail buildouts within a multi-tenant retail building. These projects will receive many shipments to the site and work being performed within their tenant space may prevent them from storing it there. Expectations must be set with the contractors at the beginning of the project to ensure all materials will be stored within their tenant space or if additional space is needed, it needs to be an approved location such as a vacant tenant space, storage room or similar space.

When projects are on a floor above grade, contractors will often utilize the exit stairs as the primary access to their floor's project. The stairs serve as a means to get workers and materials to the work areas. A common challenge is the propping of exit stair doors. These doors are often propped open to allow for the easier flow of materials and people. When these doors are left open, the stairwell's fire rating is jeopardized and the smoke control system's ability to operate successfully is also hindered.

The FPPM must ensure the means of egress are being maintained on a daily basis. Inspections conducted will identify if there are any obstructions or concerns to the means of egress system. Identified issues must be immediately reported and rectified.

Part III

Best Practices and Applications

Chapter 12

Hot Work Activities

12.1 Boston Hot Work Fire that Changed History

On March 26, 2014, in Boston, a fire broke out at a 4-story residential building in the city's Back Bay neighborhood. The fire grew out of control with the aid of strong winds on that day. This prompted a large response from the Boston Fire Department. The fire quickly spread from the basement level to the upper levels. Windows in the basement broke, allowing a large influx of air to help the fire's development. Two firefighters became trapped in the basement level and did not make it out alive.

A later investigation determined that welding activities on a nearby railing were the cause of the fire. This devastating loss of life was the catalyst for improving the city's hot work requirements. The city underwent major reform and today has a comprehensive hot work program.

I had the opportunity to sit down with John "Jack" Dempsey, retired Fire Commissioner of the Boston Fire Department, to reflect on the 2014 fire and how the city responded to hot work challenges.

Dempsey was a district chief at the time of the 2014 fire. He knew both firefighters who lost their lives that day fighting the fire. Firefighter Mike Kennedy had previously worked in Dempsey's division before being transferred to another division within the city. The fire opened a lot of eyes to hot work. That same year, Dempsey was promoted to Deputy Chief and served as the fire marshal.

There was no permit for the hot work, and there were extreme wind conditions. "Those two reasons alone should have stopped the job that day. The heavy wind coming off the Charles River blowing on the back of the building made it become like a blowtorch, and the two firefighters became trapped in the basement," Dempsey described (Figure 12.1).[1]

As a fire marshal at the time, he was tasked with combating the risks associated with hot work. He met with Jim Pauley, CEO of the National Fire Protection Association (NFPA), on developing a hot work safety training program to bring general awareness to the industry. The hot work certification program became a mandatory requirement within Boston city limits, then grew to later become a statewide requirement. He explained that "the program was not just for people doing welding

DOI: 10.1201/9781032720210-15

Figure 12.1 Firefighters respond to a nine-alarm fire at 298 Beacon St. in Boston, Mass. Photograph by Saglio, Justin. *Fire-3 (13434922053)*. BU News Service. Licensed under Creative Commons Attribution 2.0 (CC BY 2.0). https://creativecommons.org/licenses/by/2.0/.

or other hot work, it is for all construction workers." The program is designed to bring general awareness to the dangers of hot work and safe practices, so all individuals can have a basic understanding, and they can say something if an unsafe condition is observed.

This hot work training certification is a requirement for all personnel involved with the hot work operations. The individual applying for and obtaining a hot work permit must have these credentials. The individual conducting the hot work, as well as the fire watch, must also have this training. This change brought significant awareness to hot work with increased education and safeguards.

Hot work introduces a flame or fire to a building, so there is inherently a lot of risk. Dempsey explained the importance of the fire watch job function. "A fire watch can detect and suppress a fire in its incipient stage, before growing into an uncontrollable fire." He stressed the importance of having the fire watch present for the full duration of work, and after the work to prevent a fire from starting up from smoldering.

He described a firsthand experience during his time with the United States Coast Guard. Dempsey was welding a watertight door on a ship with a dedicated fire watch assigned to the operation. There was no contents within the compartment besides a small pile of material in the opposite corner. When it was suggested that they leave for lunch, Dempsey knew the hot work site must be monitored even after the stoppage of work. As he described, sure enough, 20 minutes later, the small pile far away from the welding operation caught fire. The fire was easily contained and suppressed, but the story could have been different if the scene had been left unattended.

12.2 Hot Work Introduction

When most people think of hot work, they may think of welding operations. Although this is the quintessential example of hot work, there are many construction activities that present similar risks. Hot work activities are essential on construction sites to assist with necessary work tasks. Hot work is a general term used for any work activity that produces heat, sparks or flames. Some common forms of hot work on construction projects include:

1. Welding
2. Soldering/brazing
3. Torch cutting
4. Grinding
5. Powder-actuated tools
6. Cutting metal studs or other metal materials
7. Open flames for preheating and other applications

Hot work introduces one of the required elements into the fire tetrahedron concept: heat. In addition to heat, another element, oxygen, is always present on construction sites. Therefore, the only prevention techniques against hot work fire hazards boil down to the prevention of the last critical component of the fire tetrahedron: fuel. Fuel sources would be any combustible material in close proximity to the hot work operations (Figure 12.2).

12.2.1 Common Construction Activities Involving Hot Work

1. **Welding**
   a. Structural steel connections
   b. Ductwork connections
   c. Piping connections for plumbing and gas
2. **Soldering/Brazing**
   a. Copper plumbing line connections
   b. Electrical connections
3. **Torch Cutting**
   a. Modifications to steel members

Figure 12.2 Welding Operation. Photograph by Plate Jr., William M. *GMAW welding at Dover AFB, January 12, 2004*. U.S. Air Force. Public domain.

4 **Grinding**
 a Cutting of metal components
 b Removal of material
5 **Powder-actuated tools**
 a Fastening of elements to steel or concrete
6 **Open Flames**
 a Torch for preheating steel members
 b Hot-applied roofing applications/warmers
 c Cooking equipment

A huge piece of this is education. Often, it is assumed that only welding would constitute "hot work." It is important to clearly define hot work activities and develop a hot work safety program that is easy for the project team to navigate. When hot work is properly planned, the permitting and other safeguards should not be a hindrance to the project. It is recommended that when pre-construction meetings or kickoff meetings are held for each subcontractor, hot work should be discussed. This introduces the hot work procedures early on in the process, so any impact it might have can be incorporated into the work plan.

12.3 Hot Work Prevention Plan

A key component of the construction fire safety program is a detailed hot work prevention plan. The FPPM must develop and implement a robust plan to mitigate the risks associated with hot work activities.

The hot work prevention plan should include the following process:

1 Alternative construction processes to eliminate or minimize hot work activities
2 Hot work preplanning
3 Safeguards during construction activities
4 Hot work completion of work closeout

12.3.1 Alternative Construction Processes to Eliminate or Minimize Hot Work Activities

Means and methods can vary across construction projects. Many work tasks can be completed by different approaches while still maintaining the same goal. When hot work is presented as the selected work method, it should be analyzed to determine if an alternative can be used to reduce the fire hazard risk in the building. In some instances, hot work cannot be avoided; however, it should always be reviewed to at least minimize the scope to reduce the risk of a fire.

12.3.2 Welding Alternatives

Steel connections are used for structural steel members and miscellaneous steel for stairs, handrails, mezzanines and other elements. Although welding connections create strong, fused joining of components, it is not the only way to create an adequate connection. Connections can be made by mechanical fasteners such as bolts, pins, lag screws and other nails and similar items.

Since the connection type influences the design of the various steel members, it must be considered early on in the design. In order to accommodate mechanical fasteners, pre-drilled holes, plates or other items may need to be prefabricated.

Gas piping, fuel oil piping and other piped systems often utilize welded connections. Alternative connection methods are available as described in the 12.3.3 Soldering and Brazing Alternatives section.

12.3.3 Soldering and Brazing Alternatives

Soldering and brazing are proven methods to make strong, long-lasting piping connections for the building's plumbing system. This technique has been used for many years and has become a standard plumbing practice. However, recently, alternatives have been made available that can provide a similar result. Pressed connections are a new, innovative method of joining piping. These methods include crimping pipe fittings using a specific tool designed to evenly apply the necessary pressure to create a strong seal and connection. This not only eliminates the need for soldering but also significantly improves productivity by allowing connections to be made in seconds, opposed to several minutes by traditional soldering. If the pressed connection method is acceptable by applicable codes and suitable based on the judgment of the engineer, it should be explored to eliminate hot work.

12.3.4 *Offsite Hot Work*

When hot work on building elements is necessary, it should be explored if these work tasks can be completed off-site in a machine shop or manufacturing facility. Although it seems like the same work and associated risks are just being relocated to a different building, there is more to it. Hot work and work tasks in general performed in a controlled environment, such as a machine shop, present much less risk than on a construction site. Machine shops and similar facilities are built specifically for the production of various products. Therefore, many safeguards are implemented in these shops to ensure the safety of the workers and the building. The shop workers are working under ideal conditions, such as adequate lighting, ventilation, access to proper tools and sufficient work area clearances. The shops are also provided with fire extinguishers and other fire protection systems like sprinklers and fire alarm systems. Performing high-risk activities within a controlled space minimizes risk.

12.3.5 *Mitigation Analysis Example #2:*

Activity: Installation of hot water heater.
Identified Fire Risk: Hot work associated with the soldering of copper piping to connect to the building's plumbing system. Additionally, combustible packaging and/or a wooden crate may be used for the transport of the unit.

Then the mitigation analysis is applied to this activity:

1 **Can the fire risk be eliminated altogether?** Yes.

   a As an alternative to traditional soldering, a mechanical system for joining pipes can be used that involves pressed, crimped, fittings using a specialty tool. This is commonly referred to as "ProPress" which is a brand name association.
   b The combustible cardboard packaging and the wooden crate can be removed and discarded before the hot water heater is transported into the building.

2 **Can the fire risk be reduced?** *Where elimination is achieved in Step #1, reduction analysis is not necessary.*

3 **How can the remaining fire risk be managed to prevent a fire?** After the first two steps are applied to the activity involving the delivery of appliances to the construction site, the resulting fire hazard is analyzed.

   a **Reduced Fire Hazard:** The hot water heater is installed without any hot work activity. The packaging is removed prior to bringing the unit into the building. The fire hazard is very low with this reduction.
   b **Management of Reduced Fire Hazard:**
      i Management of hot water heater installation area to ensure it does not impede means of egress routes, access to fire protection features or accessibility features.

12.3.6 Hot Work Preplanning

When hot work is necessary, it must be properly planned to ensure it will be carried out safely. Permitting processes ensure the proper procedures are in place during hot work operations.

Some jurisdictions require that a hot work permit be obtained from the fire department. Where a fire department hot work permit is required, the following steps should be followed:

1 A "Hot Work Permission Request from Owner" form (Form 12.A) must be submitted to the Owner for review and approval for each subcontractor who will be performing hot work. *The purpose of this document is to ensure the owner of the property is aware that there will be hot work activity occurring at the building. It is best practice for the owner to implement a standardized request form to ensure all pertinent information is provided, including duration, location and type of hot work. By also requiring hot work training certification numbers, you can ensure only qualified companies and workers are performing the hot work.*
2 Requests will be reviewed and approved at the sole discretion of the Owner's FPPM. If approved, the FPPM will grant an Owner's Hot Work Permission Letter to the general contractor/subcontractor (Form 12.B). *This letter serves as an acknowledgement and approval of the proposed hot work request. Some fire departments require this letter to be submitted with the hot work permit application.*
3 Contractor to apply for a hot work permit from the Fire Department using the permission letter provided. *The permitting process varies depending on location; therefore, specific permitting steps and information should be researched for the project's city/town.*
4 Once a hot work permit from the fire department has been granted, a copy must be provided to the Owner's FPPM and General Contractor's FPPM. *It is important that the greater project team, including the Owner and GC FPPM, receive a copy of the hot work permit and review it. The permit will list approved durations, locations and will often indicate if a fire watch detail through the fire department is required.*

Form 12.A

Hot Work Permission Request from Owner

This completed form must be submitted to the Landlord's FPPM. If approved, the Landlord's FPPM will issue an "Owner's Hot Work Permission Letter" to the contractor who will use the letter when applying for a hot work permit through the city of ________. NO HOT WORK MAY BEGIN PRIOR TO RECEIVING OWNER'S PERMISSION LETTER AND BEFORE THE CITY OF ________ HOT WORK PERMIT HAS BEEN SUBMITTED TO THE LANDLORD'S FPPM. ALL HOT WORK MUST BE SUPERVISED BY A FIRE WATCH (FIRE WATCH TO BE A FIRE DEPARTMENT DETAIL IF NOTED ON PERMIT).

Name of Company: ________________________________

Address: ________________________________

Contact Name: ________________________________

Phone Number: ________________________________

NFPA Hot Work Certification #: ________________________________

Location of Work: ________________________________

Address: ________________________________

General Contractor: ________________________________

Type of Space: ________________________________

Type of Work: ________________________________

Reason for the Work: ________________________________

____

____

____

Duration of Work: ________________________________

(Start date) - (End date)

Name: ____________________ Date ____________________

Company: ____________________

Signature: ____________________

Form 12.B

DATE

FIRE DEPARTMENT
ADDRESS
CITY, STATE, ZIP CODE

RE: PROJECT DETAILS
Cutting- Burning- Welding Permit
Contractor Hot Work Permit - Owner Authorization

To Whom It May Concern,

BUILDING OWNER, the owner of PROJECT NAME has engaged the general contractor listed below to perform construction for a core and shell space in a new mixed-use project at ADDRESS. The approved permit will be issued to the following contractor, for the work to be performed at the address listed, along with the specifics noted below:

Name of Company:
Address:
Contact Name:
Phone Number:
Location of Work:
Project Address:
General Contractor:
Address:
Type of Space:
Type of Work:
Reason for the Work:
Duration of Work:

The contractor/subcontractor will be required to prepare and submit an Impairment Plan as part of, and encompassing, the scope of work covered under the Hot Work Permit Application. That plan must comply with all provisions of NFPA 241. The contractor/subcontractor is required to follow any and all additional provisions or direction set forth by the Fire Department, the Fire Protection Program Manager (FPPM), the General Contractor, other designated entities, and/or personnel responsible for implementing NFPA 241 requirements. This includes, but not limited to, obtaining all associated permits and providing fire watches as deemed necessary by the Fire Department. A copy of the approved permit will be delivered and attached to the first Fire System Impairment Request and subsequent requests will reference that permit for the duration specified.

By: OWNER COMPANY
By:
Name:

Table 12.1 Hot Work Log

Date	*Time*	*General Contractor/ Tenant*	*Subcontractor*	*Type(s) of Hot Work*	*Location(s)*	*Name of Fire Watch*

In addition to the city/town permitting requirements, an internal project hot work permitting system should be developed and implemented by the general contractor.

1 A daily hot work permit must be filled out by each worker and subcontractor performing hot work. The permit must be posted at the designated project command center. The permit must include the date and time, name of contractor and worker conducting the hot work and their certificate number, the location and must be signed by the general contractor's FPPM prior to any daily hot work.
2 A daily hot work log (Table 12.1) and a fire watch log must be completed and up to date at the designated command center. Records must be maintained of these logs for the duration of the project.
3 A fire watch must be arranged at all hot work operations. A fire watch detail from the fire department is required if noted on the Hot Work Permit.
4 A Fire extinguisher must be provided at every hot work location suitable for the possible type of fire.
5 All combustible items must be removed from the work area and any area where byproducts of the hot work could land. Therefore, special precautions must be taken to protect floor openings and the edge of building areas where sparks or slag could fall. NFPA 51B, *Standard for Fire Prevention During Welding, Cutting, and Other Hot Work*, recommends removing all combustible materials within 35 feet of the hot work. Where this is not feasible, fire blankets or other noncombustible shields or barriers should be used to protect combustible items to prevent sparks from traveling from the hot work site.
6 Ensure sprinkler system is not impaired within the same zone as hot work activities. It is critical that the sprinkler system remains active; otherwise, the building is put in an extremely vulnerable situation.
7 Evaluate work zone for fire alarm devices. Smoke detectors in close proximity to hot work operations should be disabled and covered, or "bagged," to protect

the devices from dust and other debris. Impairment procedures outlined in Chapter 13 should be followed.

12.3.7 *Safeguards During Hot Work Operations*

1 Adequate Personal Protective Equipment (PPE) must be worn by all personnel involved with the hot work operation.
2 Ensure combustibles remain outside of the work zone.
3 Ensure protection deployed stays in place during the course of work.
4 The fire watch must be on-site for the entire duration of the hot work.

12.3.8 *Closeout of Hot Work Activities*

1 The fire watch must continue to monitor the work zone for a specified duration after the completion of the work. Often, 60 minutes is the required duration after the completion of the work for most types of hot work. However, in more vulnerable areas such as combustible construction or for torch-applied roofing operations, the fire watch should stay onsite for at least two hours after the completion of work. This is critical as hot spots can ignite and start a fire even after the work is completed. Applicable NFPA codes such as NFPA 1, NFPA 51B and NFPA 241 will govern this. However, other requirements should be reviewed, such as local regulations, insurance requirements and other building-specific criteria.
2 Hot work logs should be updated to indicate end times.
3 Ensure any impaired fire alarm devices are brought back into service following the impairment procedures outlined in Chapter 13.

12.4 2008 – Universal Studios Hollywood – Los Angeles County, California, U.S.A.

On June 1, 2008, Universal Studios Hollywood endured a major fire when hot work activities ignited a fire that spread across two acres of the property. Repair work was done on the roof of a building on a movie set. Torches were used for roofing work. The work was completed by 3 a.m. and by 4:43 a.m., flames were observed rising from the rooftop. The fire spread quickly across the various movie sets, including a notable New York City streetscape and a Courthouse Square (which was featured in "Back to the Future"). The fire then spread to a large warehouse. The fire continued to travel and made its way to the theme park areas.[2]

In addition to the massive property damage, fire and smoke also damaged hundreds of thousands of irreplaceable, archived film and music files. The original number of assets destroyed was pinned at 118,230; however, this number grew when extrapolated to include tallying songs on albums and different versions. It was estimated that it could have been as many as 500,000 song titles lost (Figure 12.3).[2]

Figure 12.3 *Universal Fire Smoke*. Photograph by pinguino k. October 5, 2012. Wikimedia Commons. Licensed under Creative Commons Attribution 2.0 (CC BY 2.0). License details: https://creativecommons.org/licenses/by/2.0/.

12.4.1 Reflection

This incident emphasizes the impact a small work activity, such as a roof repair, can have on not only the destruction of buildings and movie sets, but also on cherished irreplaceable archives. The impact of this fire did not just affect those directly involved with this property, but those who may have recorded music, television and movie films. Torch-applied roofing operations are especially dangerous, which is why NFPA standards require a two-hour fire watch after the completion of work.

It is important that all hot work operations are thoughtfully pre-planned leading up to the work. By securing proper approvals from owners, obtaining permits from AHJs, preparing the work area with precautions and implementing fire watches for the necessary durations, the safeguards reduce the chance of a fire occurring and reduce the potential severity.

References

1. John “Jack” Dempsey (retired Fire Commissioner of the Boston Fire Department), interview by author, October 23, 2023.
2. Jody Rosen, “The Universal Fire and the Master Recordings It Destroyed,” *The New York Times Magazine*, June 11, 2019, https://www.nytimes.com/2019/06/11/magazine/universal-fire-master-recordings.html.

Chapter 13

Fire Protection Impairments

13.1 Fire Protection Impairments

An impairment is when a fire protection system is disabled intentionally or unintentionally, and a portion of the system is out of service for a period of time. Planned impairments are a common practice in construction projects to allow for the modification of fire protection systems to accommodate the proposed layout of the space. However, if the impairments are not carried out correctly by the FPPM, the building could be in a vulnerable state due to the lack of protection. Improper impairment procedures could also result in nuisance alarms causing disruption to occupied buildings.

The management of fire protection impairments is one of the most critical responsibilities the FPPM is tasked with. Critical life safety systems must be managed by qualified individuals who understand how they operate, how they should be impaired, and how they need to be maintained for the longevity of the building.

The following are the most common types of impairments:

- **Smoke Detection** – this involves the disabling of smoke detectors by either deactivating the device on the fire alarm panel or covering the smoke detector device with a "bag," or both. Bagging a smoke detector often involves covering the detector with a plastic-fitted cover to protect the device from dust and debris from entering the chamber.
- **Automatic Sprinkler System** – this involves the disabling of the fire sprinkler system, generally by draining water from the system or zoning off sections by shutting control valves. Dry sprinkler systems, pre-action and deluge fire sprinkler systems also fall under this general category and require special impairment protocols to be followed.
- **Fire Alarm System** – this involves the disabling of fire alarm components, including initiating and notification appliances. Initiating appliances include any device that would put the fire alarm system into alarm: pull stations, heat detectors, smoke detectors, duct smoke detectors and carbon monoxide detectors. Notification appliances include fire alarm devices that provide audio and visual alarms to occupants, including horns, speakers, strobes and beacons.

DOI: 10.1201/9781032720210-16

- **Standpipe System** – this involves disabling the standpipe system. The standpipe system is a system that typically extends vertically throughout the building, often in exit stairwells, to provide hose outlets for fire department use, and sometimes is combined with the sprinkler riser to feed sprinklers on each floor of the building.
- **Other Impairment Types** – this involves the disabling of smoke control systems (stairwell/elevator pressurization systems and atrium exhaust systems), disabling of fire extinguishers, and the disabling of passive fire protection systems. This could also include the disabling of exhaust and fresh air supply fans serving parking garages. Passive fire protection systems include building elements that provide protection to occupants and the building structure. This includes spray-on fireproofing, fire barriers/walls, fire doors or means of egress systems. Any building system that functions or assists in providing a life safety purpose would be considered an impairment if the system is disabled.

13.2 Impairment Coordination

The FPPM is responsible for receiving impairment requests, coordinating the impairments with the necessary parties, ensuring the impairment is being carried out per the construction fire safety plan, and verifying the system is fully restored at the end of the impairment to ensure all systems are active.

The FPPM should develop and implement a fire protection impairment request form similar to Figure 13.1. This form includes pertinent information regarding the impairment that allows the FPPM to understand the nature of the request. This form should be issued to all contractors working in the building. The scheduled date and time of the request are reviewed against other work occurring in the building to ensure there are no overlapping impairments. Applicable codes and standards, authorities having jurisdiction (AHJs), insurers and other stakeholders can influence requirements for impairments. It is important to check with these entities to ensure that your impairment protocols are aligned with theirs. Generally, impairments must be scheduled in a manner that follows these basic guidelines:

- Only one fire protection system can be impaired in one zone at a time. For example, if the fire sprinkler system is impaired, no other system in the same zone can be impaired, including the fire alarm system or smoke control system. Furthermore, hot work activity cannot occur in this impaired zone.
- Contiguous floor levels cannot have the same fire protection system impaired simultaneously. For example, if the third level of a building has the sprinkler system impaired, the second level or fourth level cannot have the sprinkler system impaired at the same time. This would create a span of multiple floors without sprinkler coverage. In special circumstances, contiguous floors can be impaired simultaneously if a temporary fire protection system is provided to mitigate this risk. This could include a fire department detail, a temporary heat detection

Fire Protection Impairment Request

Prior to impairing any fire protection system, a fire protection impairment request must be made to the Landlord's FPPM and be approved. This completed form must be submitted at least (2) business days before the scheduled impairment.

Date of Fire Protection System Impairment: ____________________

Start Time of Impairment ________ Estimated End Time of Impairment ________

Fire Protection System Impaired:

- ☐ Smoke Detection System
- ☐ Fire Alarm System
- ☐ Automatic Sprinkler System
- ☐ Standpipe System

Extent of Impairment __________ Floor: ____________________

Reason for Impairment: ____________________

Person Conducting Work: ____________________

Contact Phone for Person Conducting Work: ____________________

Person Responsible for Mounting Impairment Tag: ____________________
(Figure 1)

Has Permit Been Pulled for Impairment if Required by Fire Department:

Yes
No
Permit Not Required

Fire Watch Required?:

Yes
No

Will Hot Work be Performed in Area of Impairment:

Yes – hot work not permitted in areas where the fire protection system is impairment unless approval from AHJ is granted and additional safety measures are taken.
No

It is the responsibility of the sub-contractor/general contractor to notify the Fire Department of all fire protection impairments. Any fees associated with false alarms that trigger the fire department to respond to the building due to improper notification will be the responsibility of the contractor conducting the impairment.

Name: ____________________ Date ____________________

Company: ____________________

Signature: ____________________

Attach insurance company impairment procedures, tag templates for valves, etc.

Figure 13.1 Automatic Sprinkler System Impairments.

Table 13.1 Impairment Log

Date	*Time Out*	*Time In*	*Level*	*Tenant*	*Contact Information*	*Contractor*	*Impairment Type*

system, or other means. This arrangement must be reviewed and approved by the FPPM, AHJ, Insurer and building owner.

- The total number of impaired floors must be in accordance with the construction fire safety plan. The FPPM, ownership and AHJ must approve this.

Ongoing impairments are logged in a master fire protection impairment log. An example of a log is seen in Table 13.1. Up-to-date records must be maintained of all fire protection impairments.

A fire watch or fire detail may be required based on the proposed scope of the impairment. If multiple systems are impaired simultaneously, the duration extends to greater than an 8-hour shift, or the extent of the impairment presents significant risks to occupants and the building, a fire watch could be required. Furthermore, insurers, authorities having jurisdiction may have specific guidelines on impairments that include fire watches.

When multiple contractors are working in a building, it is essential that the impairments are coordinated with the project-level FPPMs. A useful method is to establish a recurring fire protection impairment meeting. Upcoming impairments for each contractor can be reviewed. If impairments overlap with one another, the owner's FPPM will have to decide to accept and deny impairment requests based on priority and other factors. By carefully coordinating impairments and hot work activities, this work can be carried out without significantly diminishing the building's level of safety.

To ensure that impairments are being carried out per the construction fire safety plan, the FPPM must oversee this entire process. The FPPM must physically be located on-site so they can facilitate the impairment in a safe and efficient manner.

The sections below will outline typical impairment procedures for various life safety systems.

13.3 Smoke Detection Impairments

Dust-generating work is the most common reason for impairing smoke detectors. Sweeping, sanding, cutting and spraying are typical scopes of work that would necessitate a smoke detection impairment. Additionally, smoke detection impairments are needed when they must be relocated as part of a project.

Procedures:
Pre-planning (days before scheduled impairment)

1. Analysis of work tasks to determine the risk of activation of the smoke detector.
2. Impairment request form submitted from contractor performing work to project FPPM.
3. Project FPPM to review requests against other work occurring, extent of impairment, and duration.
4. If approved, project FPPM to schedule fire alarm service vendor to disable requested devices for the identified duration.
5. Project FPPM to notify the project team, including property managers, building engineers, superintendents, of scheduled impairment to ensure the greater team is aware.

Scheduled Impairment Day:

6. FPPM, fire alarm service vendor, building engineers and contractor to meet at the fire command center (if provided), or pre-determined location in close proximity to the fire alarm control panel at the beginning of impairment start time.
7. FPPM to verify that the correct/required devices are disabled within the work zone, working with the fire alarm service company and contractor.
8. Protective covers or "bags" to be placed on impaired smoke detectors to protect devices from damage.
9. Contractor to perform scheduled work.
10. FPPM, contractor and fire alarm service vendor to meet at the end of impairment time to discuss work status.
11. FPPM to verify with the contractor that dust-generating work is complete and the work environment is not dusty.
12. When work is complete, protective covers or "bags" are to be removed from impaired devices. FPPM to verify protective covers have been removed from all devices. *This step must occur prior to step 13 to prevent accidental activation by creating a suction when removing the cover.
13. Fire alarm service vendor to re-enable impaired smoke detection devices.

A common challenge with ongoing work requiring smoke detection impairments is that the protective covers or "bags" are not removed from the smoke detectors at the end of each working shift. This puts the building in a vulnerable state since early detection of a fire will not be possible. The risk of a fire growing to an uncontrollable size before detection increases. Therefore, it is essential to create a program that ensures that covers are removed at the end of each shift. This could include a check-out; check-in program where protective covers are tracked by unique numbers/letters. By keeping close tabs on the covers, you will know exactly what covers have been checked out and their locations. For example, the FPPM would check out smoke detector covers 1 and 2 in the morning and update the log to identify the responsible contractor, time-out, duration, and cover numbers. At the end of the shift, when the covers are returned, the log can be completed with the time in. This ensures that the covers are returned, and smoke detection coverage is not inhibited.

Some municipalities will require a smoke detection bagging permit to impair smoke detectors. The FPPM must check with local building and fire departments to understand the requirements. It is important that the project team is educated on the permitting process so that this process can be started early and not delay the project.

Some smoke detectors also function as an elevator recall feature. These are typically smoke detectors in close proximity to the elevators. If the smoke detector serving as an elevator recall is activated, the elevator will follow the approved sequence of operations, often returning to the first floor, unless the alarm is on the first floor. Since these types of smoke detectors serve additional critical safety features, they cannot be impaired without special attention and planning. AHJs and engineers should be consulted to find an acceptable solution. A potential solution would be to reprogram a pull station temporarily during construction to recall the elevators. At the completion of construction, the programming changes would be returned to the original sequence of operations.

13.4 Automatic Sprinkler System Impairments

Fire sprinkler systems are typically modified when the building undergoes a renovation to allow adequate sprinkler coverage for the new layout. In order to make modifications, the system must be disabled to drain down the system, resulting in an impairment to the system.

Procedures:

Pre-planning (days before scheduled impairment)

1 Analysis of work tasks to determine modifications required for the automatic sprinkler system.
2 Impairment request form submitted from contractor performing work to project FPPM.
3 Project FPPM to review requests against other work occurring, extent of impairment and duration.
4 If approved, project FPPM to schedule fire alarm service vendor to disable the requested sprinkler zone for the identified duration.
5 Project FPPM to notify the project team, including property managers, building engineers, superintendents, of scheduled impairment to ensure the greater team is aware.

Scheduled Impairment Day:

6 Verification of correct/required devices by the fire alarm service company and contractor to ensure associated flow switches and tamper modules are disabled for sprinklers in the work zone.
7 Fire alarm system to be put on test mode during the drain down to ensure other monitoring modules in the fire sprinkler system are not activated as part of the drain down process.
8 Fire hoses to be connected to the drain connection and extended to the permitted drain area.

9. Once the system is drained down, the system is to be put back online with the exception of the zone being worked on.
10. Contractor to perform scheduled work.
11. FPPM, contractor and fire alarm service vendor to meet at the end of impairment time to discuss work status.
12. When work is complete, the fire alarm service vendor to place the fire alarm system back in test mode to allow for the refilling of the system.
13. If a fire pump is provided in the building, power to be shut off to fire pump ahead of refilling the system to prevent the activation of the pump.
14. Sprinkler contractor to refill the system slowly to prevent any air pockets or hammering. Contractor to confirm there are no leaks in the modified system.
15. Once filled, the FPPM shall verify that the control valve is in the fully open position.
16. Fire alarm service vendor to reenable disabled sprinkler module points.

It is important to understand how the sprinkler system drains specific to your project. Often there are control valves at each floor level to zone off each floor. Therefore, this allows you to drain down a single floor while leaving the sprinkler system on all other floors active during construction operations. In multi-tenant buildings, each tenant space may be provided with its own zone control valve assembly. This allows individual tenants to drain down and modify the sprinkler system without impacting other tenants on the floor or any other areas of the building.

Some sprinkler control valves are hard-piped to a sprinkler drain that drains to the exterior of the building. Some control valves are provided with a drain connection that the sprinkler contractor will need to attach a hose to. In this arrangement, a hose is to be attached to the control valve assembly and drained to the exterior or to a container to be brought outside and drained. In any arrangement, it is critical that the sprinkler drain water is not drained onto sidewalks, patios and hardscaped areas. Water within the sprinkler system is stagnant and can become discolored as a result of corrosion. This discolored water can stain hardscape areas and can be very difficult to clean off. Furthermore, in freezing climates, the drained water can freeze on the sidewalks and create slippery conditions for pedestrians.

The FPPM is responsible for ensuring the coordination between the fire alarm service vendor and the sprinkler contractor. This is critical to ensure the system is not drained or refilled without the system being in test mode and the fire alarm technician being on standby. A common challenge arises when the sprinkler contractor assumes the system has been disabled and goes ahead and drains the system. This will set off a waterflow alarm and put the fire alarm system in an alarm state, automatically dispatching the fire department.

13.4.1 Ceiling Removal

The removal of ceiling tiles in a drop ceiling system may cause an impairment to the fire sprinkler system. The fire sprinkler system's pendent sprinkler heads depend on the continuous ceiling plane to trap smoke, having it accumulate, increasing the temperature at the ceiling plane and reaching the sprinkler's activation temperature. However, when the ceiling plane is disturbed, the hot gases will continue past the ceiling grid and accumulate at the higher ceiling level (structural floor/ceiling assembly). This will significantly delay or prevent the sprinklers from activating as designed.

When this situation is necessary to accommodate work on the construction site, the FPPM must evaluate to determine a mitigation plan. One of the following strategies should be utilized:

1. Confirm that pendent sprinkler heads provide adequate coverage. If sprinklers remain within 12" of the highest ceiling when the drop ceiling is removed, coverage may be adequate.
2. Modify to upright sprinklers during construction.
3. Perform work and reinstall the ceiling grid at the end of each shift. This is essentially carried out like a typical sprinkler impairment. Therefore, all the other rules apply (no hot work in the zone, contiguous floors cannot be impaired, etc.)

13.4.1.1 Standby Service

An approach that is sometimes used where significant construction activities are still occurring in a system with live fire protection systems is what is commonly referred to as fire alarm standby service. This approach is a little different than a standard impairment. A qualified fire alarm technician is physically stationed at the fire alarm control panel for the duration of the standby time. The fire alarm system's notification system is bypassed, and the fire alarm monitoring company is notified to disregard any signals during this time period. The fire alarm technician then monitors the fire alarm system panel for any incoming signals. If any alarm signals come in, they must be immediately investigated to determine if they are legitimate or caused due to construction activity. If legitimate, they must immediately call 9-1-1 for a rapid fire department response. If construction related, they must determine the cause and remedy the issue.

Standby service can be effective at minimizing nuisance alarms, allowing construction activities to continue without the delay in coordinating impairments; however, there are some drawbacks. First is cost, since you need a fire alarm technician physically onsite for this operation, the hourly rate, and sometimes overtime can add up fast. It's not uncommon to rack up tens of thousands of dollars each month with this approach. Also, with this approach, the standby service is often perceived as a buffer. If not managed correctly, this can be abused by not carrying out impairments properly, since the standby will always catch any alarm signals caused and will not immediately report to the fire department. Therefore, workers may take risks they otherwise would not take in a building with live fire protection systems. Smoke detection impairments may not be coordinated in areas where dust-generating work is taking place. Or sprinkler systems may be drained on a floor without coordination with the building. Therefore, it is critical to maintain order if this arrangement is chosen to ensure fire protection systems are being safely managed.

Chapter 14

Housekeeping

14.1 Housekeeping Introduction

Good housekeeping and waste management practices on a construction site create safer working conditions, increase efficiency and minimize the risk of a fire. It is critical that combustible waste is collected and removed from the construction site as soon as possible. The careful management of combustible materials onsite is imperative. It is important that the means of egress systems are maintained throughout construction without obstructions or modified components that make it unsafe. Proper fuel storage programs shall be implemented to ensure fuel is being stored in a safe manner, under maximum allowable quantities. Establishing a fire command post at the construction site will create a central location containing all pertinent fire protection information, logs, permits and contact information.

14.2 Waste Contributors

Construction projects generate a substantial amount of waste. Some of the major contributors of waste on a construction site are the following items:

Wood Framing Cut-offs – in wood-framed buildings, wood members are cut to length, and cut-offs are accumulated as the building is framed.

Wooden Crates – many building materials and equipment are delivered to the site on wooden pallets.

Cardboard Packaging – tools, building materials, plumbing, HVAC, electrical, fire protection components are delivered to the site in cardboard boxes.

Wood Flooring and Trim – flooring, trim and other specialty carpentry items generate wood scraps.

Insulation – batt, rigid and other types of insulation are abundant on construction sites to properly insulate and seal the building to meet stringent energy code requirements.

Plastics – plastic packaging, protective materials and building materials may contribute to a building's waste.

DOI: 10.1201/9781032720210-17

The production of waste varies throughout the different phases of construction. Site work and concrete foundation work will see much less waste production than during the finishes phase.

14.3 Waste Management Plan

Construction waste management is crucial to prevent hazards for both the building and its occupants. Waste is one of the most significant contributors to the combustible fuel load in buildings under construction. It is essential that expectations and waste removal requirements are clearly defined at the beginning of the project. General contractors should set guidelines for waste removal to ensure the process aligns with the construction fire safety program. Procedures and protocols should be set for the quick organization, temporary storage and removal of waste.

A common practice is for the general contractor to coordinate and provide dumpsters for the entire construction site, with individual subcontractors having access to dispose of their debris. Often, a cost-sharing agreement, such as a prorated system, is arranged. This method is preferred, so there is no excuse for the fast removal of trash from the site. Whereas, if each subcontractor is tasked with managing their own trash by either their own trash collection bin or by live-loading, this often creates a significant delay in the removal of waste.

The first step of the waste management process is the collection of waste from the source. Wood, plastic pieces, cardboard and other waste must be collected immediately. The waste must be placed in trash barrels, bags, wheeled bins or similar. The area must be cleaned of all large pieces of waste and swept clear of any smaller debris.

At this stage, the waste is contained and organized. However, if trash is accumulated and left in this arrangement, this also poses a significant fire hazard. Therefore, it is important that the next step of removal from the site occurs on a daily basis, or more often, to prevent the excessive accumulation of waste. The waste containers must be brought to the site dumpster or to a live-loaded truck to be emptied.

The location of the dumpster should be included in the pre-fire plan in the construction fire safety plan. The dumpster should be located with adequate separation from the building, fuel storage and temporary offices.

14.4 Combustible Materials

While it is not possible to eliminate all potential combustible materials from the building, it is possible to closely manage them to mitigate the risk that is controllable. Combustible materials such as wood lumber and plywood, concrete forms, flooring, insulation and similar materials must be delivered to the site to be installed within the building. The manner in which deliveries are sequenced, the storage arrangement and construction activities near this area determine the presented risk.

Consideration should be given to the timing and quantity of deliveries to the site. Minimizing the time or quantity reduces the overall risk. This needs to be balanced with the other goals, such as ensuring all materials are on-site to not delay construction progress. Therefore, this needs to be analyzed and planned early in the project and coordinated with subcontractors. Anticipated production schedules must be developed so an adequate amount of materials can be on-site for construction without excess.

Next is minimizing the risk of the materials when they are present onsite. Therefore, a thoughtful location must be utilized. A location that is sheltered from construction operations is ideal. The location should maintain all means of egress, exits and walkways. By keeping the storage away from construction operations, the impacts of hot work and other heat sources are greatly minimized. If possible, storage areas should be in either protected areas with fire sprinkler coverage or in areas that can be monitored either with a fire detection system or by visual monitoring.

14.5 Mitigation Analysis Example #1

Activity: Shipment of appliances to the construction site for the upcoming installation in a residential apartment building. Appliances will be stored within the building and staged before units are ready for installation.

Identified Fire Risk: Significant combustible cardboard packaging and polystyrene (commonly referred to as Styrofoam) are stored in a concentrated manner within the building. This creates a clustered fuel load, susceptible to fast fire growth if a fire develops.

The mitigation analysis is applied to this activity:

1. **Can the fire risk be eliminated altogether?** No appliances are needed to complete the dwelling units within the building for occupancy.
2. **Can the fire risk be reduced?** Yes. Some areas to reduce risk include:

   a. Appliances should be shipped to the site only when they are ready for installation. Reducing the overall time they are being staged onsite will significantly reduce the overall risk. Utilizing multiple smaller shipments that align with the installation schedule will cut back on the number of appliances being stored elsewhere in the building.
   b. Most of the fire risk is contained within the packaging of the appliances (cardboard and polystyrene). The packaging can be removed before entering the building for temporary storage or it can be immediately unpackaged when placed inside the building. The waste can then be quickly removed from within the building, so the appliances are not stored long-term with combustible packaging.
   c. The consolidated storage of the appliances also poses a fire risk since there is a significant risk of fire spreading to adjacent appliances should a fire occur.

Another mitigation strategy may include dividing storage areas, creating separation between appliances, distributing appliances to dwelling units ready for installation or other similar approaches.

3 **How can the remaining fire risk be managed to prevent a fire?** After the first two steps are applied to the activity involving the delivery of appliances to the construction site, the resulting fire hazard is analyzed.

   a **Reduced Fire Hazard:** Appliances are shipped to the site over the course of four (4) shipments, aligning with the installation schedule to minimize the time the units are temporarily staged in the building. Packaging is removed in the loading dock and immediately relocated to the site's dumpster outside of the building to reduce the quantity of combustible contents. Appliances are stored within an area of the building that is complete with construction to minimize the adverse effects of construction activities. Therefore, in this area, it is not anticipated that there will be any hot work activity in the vicinity, nor any obstruction of means of egress.
   b **Management of Reduced Fire Hazard:**

      i Monitoring of appliance installation and inventory to ensure onsite storage is appropriate for the anticipated installation schedule.
      ii Management of storage areas to ensure ongoing construction operations do not pose a threat to the safe storage of appliances. Hot work, temporary heaters and other sources of heat should be avoided in this area. If hot work is necessary, a separate analysis should be completed to reduce the risk, which may include protections like fire-retardant protection over appliances.
      iii Management of spacing between appliances to reduce the risk of fire spread and to allow walkways between appliances and to exits.

14.6 Flammable Gas and Fuel Storage

Flammable gases and liquids are used on the jobsite to fuel various types of equipment. Diesel is primarily used to fuel site work equipment such as excavators, front-end loaders and skid steers. Gasoline is typically used in trucks, generators and other small engine equipment. Propane is used for heaters, torches and soldering operations. Acetylene is commonly used in welding operations. Adhesives, solvents and other chemicals used in construction processes may also be flammable.

The fuel storage management plan must be included as part of the construction fire safety plan. The location must be shown on the pre-fire plan. A centralized location should be selected that is safely separated from the building, temporary structures, dumpster and smoking areas.

Some jurisdictions have specific fuel storage permitting requirements. These should be explored to ensure the fuel storage management plan aligns with city/town expectations.

It is best practice for the GC's FPPM to develop and maintain a fuel storage matrix. An example of a matrix is shown in Table 14.1. The matrix serves as an

Table 14.1 Fuel Storage Matrix Example

Project			*Project Fuel Tracking Matrix*										
Contractor	*Reason*	*Item*	*Compressed Gas*			*Liquid Fuel*			*Propane*			*Fire Department Permit*	
			# of Cylinders/ Tanks	*Size of Cylinder/ Tank (CF)*	*Total Cubic Feet*	*# of Tanks (Gal)*	*Size of Tank (Gal)*	*Total Gallons*	*# of Tanks*	*Size of Tanks (LB's)*	*Total Pounds*	*Permit# Issued*	*Maximum Allowable Quantity*
TOTAL:													

organizational tool to understand maximum allowable quantities for each type of fuel, and lists current quantities of fuel. The associated permit numbers are also referenced for each subcontractor to ensure permits are successfully obtained and valid for the proposed fuel stored.

Each fuel canister shall be adequately labeled with the following information:

- Type of fuel
- Quantity of fuel
- Contractor Name
- Contact Name
- Contactor Phone Number

Fuel storage sheds or structures should be constructed of noncombustible materials. Where large quantities or different types of fuel are present, fire-separated compartments may be appropriate for each type of fuel.

Fuel canisters and tanks must be secured. Gas cylinders should be secured with restraints such as metal chains or bands and positioned to adequately support the cylinder to prevent it from falling over.

Fuel canisters and gas cylinders must be returned to the fuel storage areas when not in use. Loose fuel canisters and gas tanks are a major fire hazard if scattered unattended in the building. It makes tracking the quantities much more difficult and increases the chance of contractors bringing in additional fuel to support their work.

The biggest challenges are getting workers to return fuel to the designated areas each day and keeping quantities below allowable limits. The FPPM must regularly monitor and enforce the fuel storage requirements. Daily inspections of the jobsite must include looking for fuel within the building, finding regular culprits, and ensuring that all are complying with the construction fire safety plan requirements. Regular audits of the fuel must be conducted to compare against the allowable limits.

Fuel needs to be removed from the construction site as soon as the contractor no longer needs the fuel. Often, gas canisters will remain onsite longer than needed, as they can become forgotten when not used on a daily basis. The FPPM must keep pressure on the subcontractors to remove fuel as soon as possible. If gas canisters or gas cylinders go unaccounted for, these must be removed from the jobsite immediately.

14.7 Permit Management

The FPPM should establish a program for managing permits for the project. There are many permits that are needed on a construction site. Many of these permits require renewal periodically. If an organized system is created, it is possible to easily track the permits and coordinate the renewals as necessary. A permit tracking board should be displayed on the construction site. It is recommended that this be at the same location as the fire command post for the project. The permit tracking board shall include the following permits:

General Permits:

- Building Permit
- Electrical Permit
- Plumbing Permit
- Mechanical/HVAC Permit

Fire Prevention Permits:

- Fire Alarm Permit
- Fire Sprinkler Permit
- Smoke Detector Bagging Permit
- Construction Fire Safety General Permit
- Hot Work Permit (from city/town)
- Dumpster Permit
- Fuel Storage Permit

Internal permitting processes should be displayed at this location as well:

- Hot work internal permitting (as discussed in Chapter 12)

14.8 Fire Command Post

The project's fire command post must be carefully planned and maintained. The command post serves as a primary meeting point for FPPMs, workers, inspectors and first responders. The command post will contain pertinent fire protection and life safety-related information.

The fire command post should include the following items:

Permitting Board:

- All associated project permits

Logs:

- Hot Work Logs
- Fire Watch Logs
- Impairment Logs

Emergency Contacts:

- FPPM and Alternate FPPM Name and Phone Number
- General Superintendent

- Police and Fire Emergency Numbers
- Utility Company's Contact Info

Project's Contact Info:

- Address
- Name
- Map

Documentation:

- Master Construction Fire Safety Plan
  - Project Specific Construction Fire Safety Plan(s) (as applicable)
- Fire Sprinkler Plans
- Fire Alarm Plans
- Smoke Control Plans
- Valve location plan for gas, electrical, fire protection shutoffs

The location of the command post should be coordinated with the fire department. In new construction projects or buildings fully renovated, the site will be under the control of the general contractor. Therefore, a centralized location such as a temporary receiving area, loading dock or hoist area may make sense as this may be a location that all individuals would pass before entering the building. In existing buildings that may be partially or fully occupied during the construction process, existing, established response points should be used. Spaces such as fire command centers, loading docks, security offices and lobbies are often utilized. They should be areas that can be directly accessed from the exterior and provided with sufficient signage.

It is the FPPM's responsibility to maintain the fire command post's contents and access throughout the project. As permits, plans, contact information, and other documentation are updated throughout the project, the command center must house the most accurate information to allow for a rapid response and effective communication in an emergency. Information such as impaired fire protection systems and hot work activities will provide first responders with critical information in a fire scenario.

The fire command post also serves as a designated meeting location. The FPPM and other designated individuals should respond to the fire command post during an emergency. These team members are responsible for implementing the procedures in the construction fire safety plan and the overall safety and wellness plan. The FPPM and other designees should develop a process for getting first responders to the fire command center as quickly as possible. One individual should locate themselves at the entrance of the jobsite by the closest main road, where it is expected the fire department will travel down. This will prevent wasted time

Figure 14.1 Example of Fire Command Post Information Organized. Image illustration by St. Germain Design Co.

trying to locate the jobsite during a critical time. This individual will guide the first responders to the fire command post, where the FPPM will inform the team of the situation. Details such as the location of the emergency or various access points will become critical. Construction site conditions are almost always in flux, so it is the FPPM's responsibility to provide accurate information based on the current site conditions (Figure 14.1).

Chapter 15

Inspections

15.1 Onsite Inspections

One of the FPPM's primary job functions is to ensure the construction fire safety plan is being carried out to completion. This is possible by the FPPM's significant onsite presence on the construction site. The FPPM's regular inspections ensure that the wavering conditions of the jobsite do not present a fire hazard risk. One of the FPPM's most useful tools is a comprehensive checklist for various fire prevention and life safety items. The checklist is specific to each building and each project within the building. The tailoring of the checklist ensures all unique fire protection features, systems and building conditions are accounted for (Table 15.1).

Table 15.1 Example of a Fire Prevention and Life Safety Inspection Checklist

Fire Prevention and Life Safety Checklist				
Project Name:				
Address:				
FPPM:				
Inspection Attendees:				
Inspection Date				
General				
Item	**Compliant**	**Non-compliant**	**Responsible Person/ Contractor**	**Comments**
No smoking onsite				
No cooking equipment onsite unless permitted in approved locations				
Temporary heaters secured and installed per listing, no combustible materials in the vicinity				

(*Continued*)

DOI: 10.1201/9781032720210-18

Table 15.1 (Continued)

Fire Prevention and Life Safety Checklist				
Signage/Postings				
Item	**Compliant**	**Non-compliant**	**Responsible Person/ Contractor**	**Comments**
"No Smoking" Signage provided throughout site				
FPPM/Emergency Site Contact Posted at Entrance Gate				
Project address posted at entrance gate				
Evacuation plans/egress routes posted throughout construction site				
Each floor provided with floor markings indicating floor/stair identifiers				
Valve Location Map				
Fire Command Post				
Item	**Compliant**	**Non-compliant**	**Responsible Person/ Contractor**	**Comments**
Documentation				
Master Construction Fire Safety Plan (up-to-date version)				
Project Specific Construction Fire Safety Plan				
Fire Sprinkler Plans				
Fire Alarm Plans				
Smoke Control Plans				
Valve Location Map				
Emergency Contacts				
FPPM: Name and Phone Number				
Alternate FPPM: Name and Phone Number				
General Superintendent: Name and Phone Number				

(Continued)

Table 15.1 (Continued)

Fire Prevention and Life Safety Checklist				
Police and Fire Emergency Numbers				
Utility Company's: Name and Phone Number				
Project's address posted				
Permits				
Hot Work Permits				
Fuel Storage Permits				
Fire Alarm Permit				
Fire Sprinkler Permit				
Dumpster Permit				
Logs				
Hot Work Log				
Fire Watch Log				
Impairment Log				
Fire Protection Systems				
Item	**Compliant**	**Non-compliant**	**Responsible Person/ Contractor**	**Comments**
Fire Extinguishers				
Fire Extinguishers located throughout construction site meeting maximum travel distances				
Fire extinguishers inspected and charged				
Properly mounted or on pedestal				
Signage provided				
Fire Sprinkler System				
Fire sprinkler system 100% active aside from scheduled impaired areas				
No obstructions impeding sprinklers				
Fire Alarm System				
Fire alarm system 100% active aside from scheduled impaired areas				

(*Continued*)

Table 15.1 (Continued)

Fire Prevention and Life Safety Checklist				
Covers/bags not on smoke detectors (unless approved by FPPM)				
No obstructions impeding fire alarm devices				
Standpipe System				
Standpipe system 100% active aside from scheduled impaired areas				
No obstructions impeding standpipe hose connections				
Temporary Standpipe During Construction (if provided)				
Standpipe provided with adequate pressure				
Hydrants				
Areas around hydrants maintained with no storage or equipment				
Signage provided				
Fire Department Connections				
Signage provided				
Areas around fire department connections with no storage or equipment				
Smoke Control Systems				
Item	**Compliant**	**Non-compliant**	**Responsible Person/ Contractor**	**Comments**
Smoke control system components maintained (fans, associated doors/ windows and other components)				

(*Continued*)

Table 15.1 (Continued)

Fire Prevention and Life Safety Checklist				
Doors into smoke protected compartments must remain closed (exit stairs, lobbies, other smoke-protected areas)				
Means of Egress				
Item	**Compliant**	**Non-compliant**	**Responsible Person/ Contractor**	**Comments**
Corridors and Aisles free of stored materials and equipment				
No storage within exit stair enclosures				
Doors closed that are part of a fire or smoke-rated assembly (exit stairs, elevator lobbies, horizontal egress, rated rooms)				
Adequate lighting in exit stairs and egress routes				
Hot Work Activity				
Item	**Compliant**	**Non-compliant**	**Responsible Person/ Contractor**	**Comments**
Approved daily hot work permit issued for work				
Hot work log updated				
Required credentials for personnel				
Fire watch provided at the hot work location				
Combustibles removed within 35 feet				
Fire blankets or other noncombustible protection used to prevent sparks or embers from traveling from hot work location				

(Continued)

Table 15.1 (Continued)

Fire Prevention and Life Safety Checklist				
Fire extinguisher provided at hot work location				
Cylinders/tanks used for hot work operations properly secured/upright				
Housekeeping				
Item	**Compliant**	**Non-compliant**	**Responsible Person/ Contractor**	**Comments**
Waste/Debris organized and removed from site on daily basis				
Materials stored in orderly manner				
Tools organized and stored properly				
Equipment stored in approved areas				
Dumpster in approved location				
Dumpsters removed from site when full / each day				
Materials kept free from areas with deployable fire or smoke curtains				
Fuel Storage				
Signage Provided				
Fuel stored within approved containers or structures				
Fuel labeled				
Fuel permits associated with storage posted				
Fuel stored is below allowable limits				
Site Access / Fire Department Access				
Item	**Compliant**	**Non-compliant**	**Responsible Person/ Contractor**	**Comments**
Adequate signage provided for construction site				

(Continued)

Table 15.1 (Continued)

Fire Prevention and Life Safety Checklist				
Adequate lighting provided				
Fire department emergency access lanes maintained				
Construction hoist/ elevator access maintained				
Emergency key box provided to the site for gate and other locked areas or equipment (hoist, stairs, rooms)				
Muster location maintained				
Site fence in sound condition				

15.2 Managing Deficient Items

Deficient items must be identified and addressed in an efficient manner. The FPPM must have the authority to correct unsafe conditions. The FPPM must notify the responsible party for the deficient condition observed. Hazards that present immediate harm or injury must be rectified as quickly as possible. Based on the severity of the observed hazard, direct communication to the worker or responsible supervisor must be made immediately. Other minor noncompliant items can be summarized in the inspection report and addressed in a standardized process.

15.3 Inspection Record Keeping

Inspection records should be maintained by the FPPM for the duration of the project. These should be in both electronic format as well as hard copies maintained in a binder that can be made available to the AHJ.

15.4 Fire Department Inspections

The fire department may have an on-site presence throughout the construction phase of the project. In some cases, there may be milestones that require the fire department to visit the site. In other cases, it may be random, unscheduled visits to ensure site conditions are meeting expectations. In either case, the FPPM should coordinate the walk-through of the site. The FPPM should take detailed notes of the fire department's observations and recommendations. The FPPM is the best person to escort the fire department through the construction site since this individual has the most knowledge of relevant fire protection information on the project.

A proactive collaboration approach with the fire department goes a long way in creating a strong working relationship. The fire department should be consulted on site access, location of fire protection system components and other elements that impact the response during an emergency. This ensures all necessary components are accounted for and that the fire department is made aware of site-specific conditions. It is in everyone's best interest to invite the fire department to the construction site periodically, so they are aware of the site's changing conditions.

15.5 Firestopping Inspections

It is critical that all penetrations through fire and smoke-rated assemblies are adequately firestopped and sealed to maintain the rating of the elements. Unsealed holes, annular spaces around pipes, edge of slab joints and top of wall joints are all susceptible areas where fire and smoke can easily spread if not sealed.

Some projects may be subject to 3rd party firestopping inspections conducted by a qualified agency based on the adopted code. High-rise and healthcare buildings are often types of buildings where a 3rd party may be required. Projects subject to the 3rd party inspection requirement must meet the applicable testing standards by completing a minimum percentage of destructive tests or witnessing the installation of firestopping systems. Regardless of whether a 3rd party agency is required, it is critical that all penetrations are adequately sealed to maintain the integrity of the various fire and smoke-rated assemblies.

The general contractor shall develop and implement a quality control and quality assurance system related to firestopping systems. In some instances, there may be a single contractor responsible for firestopping all penetrations and joints across all of the trades. In other cases, each individual subcontractor may be responsible for their own work's penetrations and joints.

In the early stages of the project, the listed firestopping details should be submitted for each type of penetrating material, assembly penetrated, edge of slab condition and top of wall joint conditions. These details must be reviewed and approved by the design team to ensure the listed details maintain the necessary ratings of the elements. It may be necessary to engage a firestopping specialist in certain cases to develop engineering judgments where a standard detail doesn't perfectly align with the building conditions.

When the project reaches the stage where firestopping work will take place, a kickoff meeting should be held with the project team to review expectations. The general contractor, subcontractor representative, owner, design team and 3rd party firestopping inspector, if required, should be in attendance. The project team should review the following items in the kickoff meeting:

1 Procedures for requesting destructive/witnessing inspections (area, type of system, responsible subcontractor, date, time).

2 Procedures for documenting inspections, deficiencies, and areas closed out.
3 Procedures for visually inspecting all penetrations/joints on each floor to ensure 100% of all penetrations and joints are adequately fire- and smoke-sealed.

When there are multiple general contractors working in the building or vendors engaged directly for the owner, it is critical that they develop and implement a firestopping inspection program as well. Companies such as internet providers, security companies and beverage companies may need to run wires, lines and other components through fire- and smoke-rated elements; therefore, they should also ensure their work maintains the ratings needed.

15.6 Flammability Testing and Certification

Upholstered furniture, mattresses, decorative materials, draperies, wall coverings and other elements may be subject to flammability standards to ensure these elements do not present an additional fire load in occupied buildings. Adopted fire and building codes will indicate which items are regulated and what standard needs to be met. California Technical Bulletin 117 or better known as CAL 117, sets flammability standards for upholstered furniture. Although originally enacted in California, these requirements have since expanded much further across the United States and other countries. ASTM International produces standards that are often adopted by building and fire codes specific to the flammability of certain materials. It is essential to determine what standards are applicable, what building furnishings and elements are regulated, and to specify items that comply with these requirements. Some jurisdictions may also require the submission of the flammability certificates, indicating compliance with the required standards. Some jurisdictions may even require their own lab testing of materials where different testing standards are established.

While this is mostly enforced during the initial occupancy of the building or renovation project, any regulated building element must meet the applicable requirements. Therefore, it is important that property managers and building engineers understand these requirements to ensure long-term compliance with these standards in the building.

Chapter 16

Fire Department and Authorities Having Jurisdiction (AHJ) Collaboration

16.1 Critical Collaboration

The collaboration between the project team and the building and fire departments is often overlooked on construction projects. There is sometimes a stigma surrounding this. The authorities having jurisdiction (AHJs) can be seen as the rule enforcers, prohibiting maximum productivity by imposing permitting and other requirements on the project. From the AHJ's perspective, the project team can be seen as having little regard for fire prevention or the safety of first responders and only focusing on money and schedule. However, a strong relationship can and needs to exist between the project team and AHJ for a successful and safe project.

The partnership with the building and fire departments must happen early in the project, in the planning stages, before the commencement of construction. For new construction or large, complicated, challenging projects, a meeting with the AHJ is recommended to review the project. This meeting introduces the project to the building and fire departments. The building's design and fire protection features are reviewed. Details such as occupancy type, construction type and site layout are presented. Any initial concerns or recommendations can be discussed. Updates to the plans can be made at this stage without significant project impacts.

16.2 Initial Presentation of Project

This initial presentation of the project provides the following benefits:

- Introduction of project team (Owner/Developer, General Contractor, Design Team) with AHJ (Building/Fire/Zoning Departments). It is important that personal relationships are established. At the end of the day, all concerns, issues, recommendations and requests are made person to person. Decisions could be subconsciously influenced by personal feelings about a particular person/company.
- Advancement of permit – the issuance of the building permit and other supplemental permits is critical on a construction project. The delay of having a permit in hand could be costly to the project. The open review and discussion with the

DOI: 10.1201/9781032720210-19

AHJ and the project team will help answer questions that might otherwise be received much later after the formal permit review comments are sent back to the team.

- AHJ recommendations – recommendations such as location of fire command centers, fire department connections, fire pump rooms and other elements that require the collaboration with the fire department can be understood. The recommendations can be incorporated into the plans before it becomes harder to make significant changes to the project.
- Presentation of construction fire safety approach – the contents of the construction fire safety plan can be reviewed. The pre-incident plan should be reviewed in detail to inform the fire department on the location of pertinent life safety elements and site access.

16.3 Ongoing Collaboration with the Fire Department

Design changes, building or fire code variances should be reviewed with the AHJ before any submissions are made. By seeking the AHJ's opinion and working with them, they will feel much more included in the project. The support of the AHJ on unique situations is essential for favorable outcomes when items go to hearings or appeals.

The fire department should be provided with all updates to the construction fire safety plan. It is important that they are provided with the most accurate information, especially on a rapidly changing construction site. The project team should invite the fire department out to the site after each significant change to the site. Major project milestones may include the evolution from excavation to foundation work to structural work to MEP/FP install and to finishes.

Meetings may be warranted for specific fire protection and life safety systems. For example, the submission of fire alarm plans or sprinkler plans for the fire department for a permit may initiate a meeting with the project team. Rationale, design approach and unique conditions can be reviewed and can usually be communicated best by in-person meetings. It is usually helpful to start with an overview of the project and to present a few floor plans to properly orient the fire department to the project.

16.4 Fire Commissioner's Perspective

When I discussed the collaboration between the fire department and the project team with Jack Dempsey, retired Fire Commissioner of Boston, he emphasized the importance of this critical relationship. He reflected on his 36-year career with the Boston Fire Department. He served multiple roles, including fire marshal and ultimately fire commissioner. He described how it is essential that there is a level of trust between the fire department and the project team and ongoing cooperation. He explained that the fire department is not there to shut a job down but rather partner with the project team to meet both organizations' goals. He highlighted the significance of strong communication to ensure site conditions are understood by the fire department. He said, "life safety is our number one goal, both the lives of construction workers and firefighters fall under that."[1]

Dempsey recommended having the fire department perform site visits at major milestones of the construction project to familiarize the fire department with current conditions. Dempsey stressed the importance of taking a proactive stance on construction fire safety: "Emergencies are stressful, anything we can do ahead of time to take some of the stress away is helpful for everyone."

"Construction site fires are dangerous period. Due to a lack of systems in place and the delay in the notification of the fire department." Jack Dempsey explained the vulnerabilities of buildings under construction. "Fire protection systems are incomplete. In wood construction you don't have plaster on the wall to protect the structure. Hazards such as hot work, waste, smoking continue to be common hazards."

He described a fire that occurred in Dorchester, a neighborhood in Boston, that tore through a building under construction in 2017. An 83-unit residential building was nearing its final completion when a fire started on the roof of the wood-framed building. Inactive fire protection systems made the fire especially hard to contain. Dempsey described challenges with this fire, including that the fire standpipe had a valve closed, preventing access to firefighting water during the fire. This and the other challenges associated with lightweight wood-framed construction made fighting the fire very difficult. The fire had to be fought from the exterior of the building only. This fire and other similar incidents highlight the importance of the coordination required between the project team and the fire department.

16.5 Training Opportunities

If there are training opportunities that could be offered to the fire department, these should be considered. There is an obvious opportunity if the entirety of the building is being demolished; the fire department can use the building for a number of training exercises. However, smaller items, but just as meaningful, could be offered. For example, during the excavation and foundation phase, the fire department may want to utilize the crane to extract a stretcher with a dummy. This would mimic techniques used to quickly get an injured construction worker out of hard-to-access areas. If there are materials that are being disposed of, such as glass, brick or concrete, they may have some use for these items as part of their training. These have little to no impact on the construction project, but can provide huge benefits to the fire department.

Reference

1. John "Jack" Dempsey (retired Fire Commissioner of the Boston Fire Department), interview by author, October 23, 2023.

Chapter 17

Challenges and Effective Practices

17.1 Importance of Effective Fire Prevention Plan

The effectiveness of the construction fire safety program is directly correlated to the priority it is given on the construction project. A strong safety culture must be formed on the construction site with a major emphasis on fire safety. The success of the plan is a function of the development and implementation of the program.

It is essential that a high-quality construction fire safety plan is created for the project. The plan must be specific to the project and address the hazards that are unique to the project. A common challenge is that the construction fire safety plans can be seen as a checkbox to satisfy a city or town permitting system. In some cases, very little effort or time goes into producing a useful plan. Plans are developed based on carryover information from a previous project. Critical information is missed, and irrelevant information gets included.

Equally important, if not more, is the implementation of the construction fire safety program. High-quality plans that do not get implemented are just as productive as having no plan at all. Often, the CFSP is developed in the early planning stages of the project. Assumptions are made at that time to finalize the plan. The designated FPPM coverage, project durations, phasing, expected hot work and impairments are projected based on this early analysis of the project. Many changes can occur from this point to when the project actually starts.

Modifications to staffing plans are a common evolving component of construction projects. The precise commencement of the project may prohibit someone who was planning on being on the project from being there due to existing commitments on other projects. Different construction methods may be used after consulting with the subcontractors on the project. The current need for impairments or hot work may be different than what was originally assumed. Everyone in the construction industry knows that these types of changes are common. Since changes cannot be stopped, the solution is to adapt the CFSP to the constantly changing project.

The CFSP must be revisited prior to the commencement of construction to ensure all contents are still accurate. Although the FPPM and the alternative FPPM are primarily responsible for managing the CFSP, the broad team must understand the program as well. A construction fire safety review meeting at the start of the

DOI: 10.1201/9781032720210-20

project is highly recommended. The master CFSP, project-specific CFSPs, impairment procedures and hot work procedures should be reviewed to ensure the team understands items that may impact their job function. This allows all team members to properly inform trade partners to ensure compliance with the overall program. For example, properly planned hot work tasks will not interfere with productivity. However, if the contractor is late applying for a permit, or did not arrange for a fire watch, or did not coordinate other work activities near the hot work properly, then project delays or other hazards could present themselves.

17.2 Common Challenges

Additional challenges can be seen on smaller construction projects where the sum of all field coverage is by a single person, a superintendent. The superintendent wears many hats in this scenario. They are the primary supervisors of construction, sometimes manage RFIs/submittals, manage material deliveries, and are responsible for general safety on the jobsite. This individual may also be appointed to be the FPPM. From a staffing and financial planning perspective, it makes a lot of sense to have the individual already on-site also act as the FPPM. However, there can be conflicting goals when this dual role exists. Superintendents are mainly focused on schedule and cost. They drive work forward, deal with issues, resolve them and keep moving forward. Some dual role superintendents may see the FPPM roles and responsibilities as a hindrance to their productivity and the CFSP may be put on a back burner. These are the projects that pose the most risk. When there is no regard for construction fire safety from the designated FPPM, there will be no construction fire safety culture fostered. There is subsequently a trickle-down effect to the subcontractors and individual workers. Unsafe work activities will go unnoticed or disregarded. This is the importance of keeping the CFSP top priority at each level of the project: ownership, general contractor, subcontractor and individual workers. Without full commitment, the program will not be successful.

Another common challenge is compromising construction fire safety for schedule or cost benefits. In mismanaged jobs, sometimes corners can be cut to make up for improperly planned tasks. For example, if a worker has only one day's worth of hot work over several months on the project, they may see the risk associated with this as low. By not going through the proper channels, notification was not made to the owner of the building, a city/town permit was not received for the hot work, a fire watch was not arranged to monitor the work, credentials were not verified for the personnel involved, and logs were not maintained. This unpermitted work can present a significant risk to the project. Fire protection systems may be impaired in the same zone where the hot work is taking place. A fire that may otherwise be quickly suppressed by the sprinkler system or by a fire watch may grow to a much larger and impactful fire. Other work activities or material deliveries may also present a risk with the unpermitted hot work. If combustible materials or waste are near the hot work, the management of an otherwise simple hazard becomes much more difficult.

When a strong construction fire safety culture is established on the construction site, these types of unsafe activities should be noticed quickly and stopped. It is only when each individual worker, along with the entire project team, prioritizes construction fire safety that the CFSP will be extremely impactful.

17.3 Overcoming Challenges

17.3.1 Designated Construction Fire Safety Meetings

Establishing kickoff meetings and other recurring meetings throughout the project can overcome some of the challenges. Specific meetings on fire safety will educate the greater project team and will also keep the principles of the plan fresh in mind.

A kickoff meeting should be held at the commencement of construction to review the CFSP. Recurring fire prevention meetings should be held throughout the project. A weekly cadence should be used as a starting point, and it can be adjusted from there based on the project's needs. The following items should be reviewed at each meeting:

1. Upcoming fire protection impairments
2. Upcoming hot work activities
3. Upcoming fire alarm testing
4. Upcoming modifications to means of egress
5. Upcoming site access changes
6. Fire Department inspections and acceptance testing

If more than one project is occurring in the building, each project's FPPM must attend the meeting. For example, in a mixed-use building, there may be an office build-out on level 2 and a retail build-out on level 1. In this scenario, the proper coordination is even more crucial to prevent overlapping impairments or hot work activities. The common denominator in these arrangements is the owner/developer/building manager. The two individual projects (office and retail tenants) are solely focused on their project and space within a vacuum. Therefore, the owner, or the owner's representative, must be the primary person managing the construction fire safety program.

Chapter 18

Construction Site Hazards

18.1 2017 – 264 Unit Residential Building – Waltham, Massachusetts, U.S.A.

In 2017, a large residential apartment complex under construction in Waltham, Massachusetts, a suburb of Boston, suffered a fire that destroyed most of the 264-unit complex. The fire broke out in the early morning hours on Sunday, July 23, 2017. Five buildings that made up the new development were completely destroyed. The damage from the fire was estimated at $110 million. Investigations determined the cause of the fire to be arson. The majority of the development primarily consisted of wood-framed structures. The buildings were expected to open the following year.

Several factors played into how the fire transpired and its magnitude. The buildings were well underway with construction over their multi-year schedule. The buildings were at an extremely vulnerable point at the particular stage of construction. The structure of the wood-framed buildings was complete and enclosed. Fire protection systems were not yet activated and protective elements for the wood structure were not complete. Lacking site security and control also contributed to the events. Security guard service was not in place.[1]

At the time of the fire, I was living in Newton, a neighboring city to Waltham. When the news broke about the fire on the morning of July 23, 2017, I visited the scene. The project was only a few minutes from where I was living. As I approached the scene, the streets were closed, and police were directing traffic. I found a spot to park my truck and walked towards the scene. The smell of charred wood wafted through the air. I saw a massive construction site that was completely obliterated by fire, with only a few distinguishable elements, such as concrete elements and a parking garage structure. The fire department still had a major presence on the site. They were continuing to spray water on the rubble, as hot spots took hours to cool (Figure 18.1).

18.1.1 Reflection

The risk of arson in buildings under construction is particularly high, with an even higher severity level. Access into the construction site and building in the case of the Waltham fire was not secure enough to deter a trespasser. A secure perimeter fence

DOI: 10.1201/9781032720210-21

Figure 18.1 Destruction of Cooper Street Development. Photograph by Collier, Jon. Wikimedia Commons. https://commons.wikimedia.org/wiki/File:Destruction_of_Cooper_Street_Development.jpg. CC BY-SA 4.0.

with abundant lighting is your first round of defense. However, it should be expected that this layer could be breached, so additional safeguards must be put in place. The next is securing the building itself. It may not be possible to completely secure the building at the early stages of the construction, but as the construction progresses, the securing of the building must be prioritized. When the building structure is complete, at least the first-floor windows and doors should be completed as well. Even if someone were to break into the site, things such as a tight perimeter fence, lighting, a secure building with locked windows and doors will slow down access to the building, which allows additional time for the intruder to be detected.

Additionally, a means of monitoring the site should be implemented at appropriate times during construction. While it is likely not necessary during site work and excavation, it will be extremely valuable when the structure is far along and before life safety systems are commissioned. Video monitoring or an on-site guard must be considered.

The installation of a temporary fire detection system can also prove to be very useful. This allows the fire to be detected at an early stage. Notification can then be made to a monitoring company, fire department or designated parties to investigate the fire before it grows to an uncontrollable stage.

In the case of the Waltham fire, the arsonist was provided access to the site without detection. A fire was started in the building without detection. Notification to the fire department was delayed due to the lack of monitoring or fire detection systems. When authorities arrived on-site, the fire had developed to a point that no firefighting efforts could control. This culmination of breakdowns led to the severity of the fire.

18.2 Site Security

A security program must be implemented to protect against trespassers on construction sites. This is important for theft, vandalism, liability and for fire prevention purposes. If these buildings are left unmonitored and unsecured, the risk of a fire and the severity of that fire are substantial. Vacant buildings, including those under construction, garner attention from arsonists. Particularly, large wood-framed buildings are extremely susceptible to intentionally lit fires due to the lack of active fire protection systems, exposed wood structure and no compartmentalization in place.

Construction sites must have a continuous fence around the perimeter of the site without any gaps or holes. While perimeter fences will keep ordinary people out, they will do little to keep someone out who has a strong desire to get in. Therefore, additional protection and monitoring are necessary.

There are several options when it comes to finding solutions to monitoring the construction site after hours. A traditional arrangement may include guard service by security personnel who patrol the site and perimeter. While this may successfully secure the site, it may be cost-prohibitive for some projects. There are also many technologies offered that can provide video surveillance monitoring in real-time. Cameras can be placed around the construction site and in the building as well. These cameras are monitored remotely off-site, so no on-site presence is necessary. When motion or suspicious activity is observed, alarms are activated and local law enforcement is called. It is important that the perimeter of the site is well-lit, as this allows security guards and cameras to detect motion or breaches into the site with ease.

New buildings under construction also do not have the benefit of quickly detecting the fire and dispatching the fire department. Small fires that may otherwise be able to be quickly extinguished by onsite personnel may grow to a large, uncontrollable size before they are detected or before the fire department arrives. A temporary fire detection system should be considered. Smoke or heat detectors can provide some level of early detection and notification. These systems could play a major role in catching the fire in its early stages, rather than detecting it too late and having a total loss of property.

18.3 Smoking

Smoking must be prohibited from the jobsite. Smoking in any form (cigarettes, cigars, pipes and similar items) presents a significant risk to the building. The small, smoldering characteristics of smoking items make them especially dangerous. If cigarettes or similar items are discarded into waste containers, barrels or dumpsters, they can ignite hours later when the construction site is unmonitored and grow into a much larger fire. There must be a 100% ban on smoking everywhere on the construction site. "No smoking" signage must be provided at the perimeter of the project site as well as dispersed throughout the site. Smoking receptacles should be provided at the main entrance to the construction site to allow workers to safely discard, extinguish and contain smoking items.

Smoking must be closely monitored and enforced on the construction site by the FPPM. Workers may be enticed to smoke discreetly if accessing acceptable locations is difficult or time-consuming. In high-rise buildings, often hoists or temporary elevators are busy throughout the day, bringing materials and workers up and down. So, when a worker is presented with waiting 15 minutes or more for the construction hoist or traveling down several flights of stairs, they may decide to take a chance and smoke within the building. The FPPM must have a protocol in place if smoking is observed. Typically, a one-time warning for the first offense and safety orientation is common. Second offenses or more should be handled by removing the worker from the job site. If consequences are not strictly enforced, there will continue to be frequent offenders.

Vaping and e-cigarettes should fall under the same umbrella. While the risk is slightly different since there are batteries involved, as discussed in the lithium-ion battery section, there is still a risk of fire. Many jurisdictions have smoking bans in workplaces that would further prevent smoking on the job site.

18.4 Temporary Heating Equipment

Some climates warrant temporary heating equipment during some parts of the year. Maintaining a minimum temperature within the work area is necessary to create a comfortable work environment and also to continue with construction operations that are subject to minimum temperatures. Concrete, adhesives and roofing applications all have minimum temperature conditions that must be met. Therefore, heating equipment is often unavoidable.

In existing buildings, the building's HVAC system should be utilized to provide heat to the construction area. This will prevent the need to bring in a portable heating solution that would present different fire hazards. The use of unit heaters is a good solution to provide heat in a construction zone.

There are several different options and types of equipment that can provide temporary heat to the construction site:

18.4.1 Direct-fired Heaters

This type of heater uses a fossil fuel to create a flame to heat the surrounding area. These are often referred to as "salamanders." These are popular due to the portability of moving them around as needed and the low cost of purchase. However, these types of heaters present the most risk. The movability of these heaters, if not managed correctly, can present a major fire risk.

The heaters must be situated on a noncombustible, level surface. Sufficient clearance must be maintained around the heater to walls, equipment, waste and other combustible materials. Adequate ventilation is critical since these heaters will use much of the available air in the space and will produce toxic byproducts of combustion. Therefore, ventilation must be considered and monitoring devices should be provided to detect carbon monoxide and nitrogen dioxide.

18.4.2 Indirect-fired Heaters

This type of heating system operates similarly to a house ducted heating system. A heating system utilizing fossil fuel is located outside the building. A heat exchanger warms the outside air and is distributed through the building by fans, ductwork and diffusers. A benefit to this arrangement is that outside air is used for combustion and byproducts of combustion are exhausted outside. Therefore, only clean heated air is distributed throughout the building.

18.4.3 Electric Heaters

This type of heater uses electricity from a battery or temporary power source to create heat. Ducted systems or radiant heating systems can be electric. The advantage of this type of heating is the clean emissions compared to toxic gases from the byproducts of combustion when fossil fuels are used. However, there are limitations to these products, such as the available electricity on the jobsite that can support the energy use of the heaters.

18.5 Combustion Engine Equipment

Combustion engines are commonly used for several construction-related equipment. Machinery, generators, concrete power floats and other tools may use combustion engines. The surrounding environment must be evaluated to ensure proper ventilation is provided. Clearances must be provided around hot elements on the equipment. If combustible materials or waste encroach on the space surrounding the equipment, it could create an ignition source that could ignite the adjacent materials.

Where possible, combustion engine-powered equipment should be placed outside, exterior to the building, with sufficient clearances. When the equipment must be located inside, the space must be assessed for adequate ventilation. Natural and mechanical ventilation options must be considered. Doors, windows, and other openings to the exterior can be used. Temporary exhaust ventilation can be used as well.

18.6 Carbon Monoxide Monitoring

A carbon monoxide (CO) monitoring program must be developed if fossil fuel-powered heaters or other equipment utilizing internal combustion engines are present onsite in enclosed or partially enclosed areas. The monitoring program should include several battery backup CO detectors located throughout the building where the equipment is present and other atmospherically connected areas. In addition, CO detectors with parts per million display readings should be utilized to track fluctuations in CO concentrations. This will ensure the levels of toxic gases from the combustion engines are maintained below acceptable, safe levels. These concentrations should be recorded in a log. Protocols must be developed so that if levels are above normal, procedures are in place to introduce more ventilation and immediately stop the use of the equipment. A threshold of immediate evacuation must be set and understood to ensure that if levels of CO get to a dangerous level that the area is evacuated and ventilated before any work can continue.

18.7 Cooking Equipment

According to a study by NFPA, the leading cause of fires on construction sites is from cooking equipment.[2] Therefore, consideration must be given to cooking operations on a construction site and protocols must be outlined in the construction fire safety plan. Cooking operations should be prohibited altogether on the jobsite. Workers should not be permitted to bring grills, griddles, microwave ovens, hot plates or other cooking equipment to the site. Cooking equipment can present unwanted fuel and heat sources to the construction site. Open flame burners, propane stoves and charcoal grills all create significant fire hazards on the construction site. When these operations take place on a construction site, the environmental factors create a much higher hazard than if this equipment were used elsewhere. For example, the discarding of used charcoal left on the construction site could ignite hours later and develop into a larger fire. Other hot elements from cooking equipment can also be an ignition source, both when the equipment is in operation and for some time after.

If cooking operations on the construction site are necessary, they should be worked into the overall construction fire safety approach. Designated areas, required signage, and protocols should all be outlined in the report. Less hazardous appliances, such as microwaves, should be proposed before allowing more hazardous appliances. Microwave ovens should only be proposed where they can be used in a manner that meets the manufacturer's requirements. Grills, burners and griddles using propane must be used in a remote location on the site, away from the building and other structures or combustible materials. Adequate signage must be provided to designate cooking areas. Charcoal and other solid fuel cooking equipment should be prohibited from construction sites due to the extended cool-down phase after use. Designated cooking locations must be indicated on the pre-fire plan and strictly adhered to.

18.8 Lithium-ion Batteries

There has been much focus recently on the fire hazards associated with lithium-ion batteries. Lithium-ion batteries are finding their way into vehicles, power tools, powered scooters, bicycles and other handheld electronic devices. They are desirable due to their high energy density, providing a significant amount of energy at a relatively low weight. These types of batteries have several benefits, including long lifespans, fast charging times and other positive qualities. However, conversely, these batteries can present a unique fire hazard that poses many challenges. When lithium-ion batteries are involved in a fire, typical extinguishment methods may not be as effective. The batteries can enter "thermal runaway," making firefighting efforts extremely difficult. Thermal runaway is described in the Encyclopedia of Electrochemical Power Sources as:

> *A thermal runaway is an incident where one exothermal process triggers other processes, finally resulting in an uncontrollable increase in temperature. This can result in the destruction of the battery or, in severe cases, also in fire.*[3]

Lithium-ion batteries that are damaged, charged incorrectly, or overheated are susceptible to developing a fire. Therefore, the construction fire safety plan must also address the proper management of lithium-ion tools, equipment, scooters and bicycles.

Workers on projects located within large cities may utilize powered scooters, bicycles and similar transportation types to commute to the construction site. Limited parking, high parking costs or other convenience reasons may lure workers to utilize alternative transportation options. Designated parking areas must be established on the construction site to safely store these. These parking areas must be incorporated into the construction fire safety plan and shown on the pre-fire plan to ensure that the location does not present a risk. Areas that provide adequate separation from the building, fuel storage locations, temporary field offices and other combustible materials or waste should be selected. Designated signage should be provided, and the FPPM should monitor to ensure that no EV transportation vehicles are located within the building or adjacent to hazardous items.

The management of lithium-ion power tools should also be considered as part of the construction fire safety plan. When lithium-ion batteries of power tools are cared for and charged following the manufacturer's recommendations, very little fire risks exist. However, when tools are dropped, hit or otherwise used against the manufacturer's guidance, they can become damaged. Tools showing evidence of damage should be replaced and removed from the jobsite. The FPPM must closely monitor this with the assistance of other supervisors on the site.

Lithium-ion batteries for power tools are often interchangeable between several different kinds of tools. Therefore, it is not uncommon to have many batteries on a jobsite. Large sites could have upwards of 100 batteries between several subcontractors. Holding subcontractors and their workers accountable for the safe management of the lithium-ion batteries is very important. Ensuring only listed and approved chargers and batteries are in use is critical. The FPPM must monitor and enforce the safe management of batteries on the construction site to minimize the risk of a fire.

References

1. Logan, Tim, and Travis Anderson. "Waltham Apartment Complex Fire Was Arson." *Boston Globe*, August 10, 2017.
2. Campbell, Richard. *Fires in Structures Under Construction*. Quincy, MA: National Fire Protection Association, October 2023. https://www.nfpa.org/education-and-research/research/nfpa-research/fire-statistical-reports/fires-in-structures-under-construction.
3. Sauer, D. U., "Secondary Batteries – Lead–Acid Systems | Lifetime Determining Processes," in *Encyclopedia of Electrochemical Power Sources*, ed. J. Garche (Elsevier, 2009), 1017–1036.

Chapter 19

Transition from Construction to an Occupiable Building

19.1 Construction Phase to Operational Building

When you think of a building's life cycle, it has many phases. Before the building even exists, the building undergoes massive planning and designing efforts. Next, the construction phase is a crucial phase in the building's life. Then the building is complete and becomes an occupiable building. Particularly in new construction, there are several challenges to overcome in this transition from construction to completion. The final stages of construction include livening the building systems, including fire protection systems. Once these systems are tested and accepted by the AHJ, engineers and other regulatory agencies, these systems must be maintained and carefully managed going forward.

There is a major shift in how the building operates at the tail end of the project. Workers and supervisors are used to the building being under construction with no live fire protection systems and under their full control. If a fire alarm device needs to be moved to accommodate a work activity, it can be moved without a major consequence. Modifications to the fire sprinkler system can be made without draining the system, disabling fire alarm points and carefully carrying out the impairment process. This all changes when the systems are live. This usually happens while there is still a significant amount of work that needs to be completed. Or in some cases, there could be other "Day 2" projects or tenant improvement work occurring. Managing dust-generating work, fire sprinkler shutdowns and maintaining egress routes becomes the utmost important.

In multi-tenant and core and shell construction, the completion of the base building work is not the end of construction, but in some ways, the start. When the project reaches substantial completion, temporary certificate of occupancy (TCO) or another similar milestone, often, tenant improvement (TI) projects are just kicking off. This is when the owner must step up in the FPPM role and manage the entirety of the construction fire safety program across all active projects occurring in the building. It is not possible to hand this responsibility to the general contractors working on the TI projects. The general contractor performing the TI work is likely working in a limited zone. In the case of a retail build-out, it may consist of 5,000 square feet of the first floor within a 30,000 square feet floor footprint. Or it could

DOI: 10.1201/9781032720210-22

include the office buildout of an entire floor of a 10-story building. In either case, the general contractors working on these TI projects are solely focused on their individual projects and not considering the impact they may have on the overall building or other projects taking place in the building. These TI projects are separated from outside influences. Meaning they are fully engulfed in their projects with their top priority being to maintain schedule and budget without considering the impacts their project may have externally. The tenant GC may have a tendency to compromise safety at the expense of the life safety of the building. It is the responsibility of the owner's FPPM to hold the project-level FPPMs accountable. Each project's presented risks must be closely managed to not jeopardize the status quo of the building's life safety.

These TI projects will require fire protection impairments, hot work activities, and possibly means of egress encroachments. If the owner is not carefully reviewing, approving and managing the overall construction fire safety program, the building could be put in an extremely vulnerable state. If a TI project on the second floor of a building is impairing the sprinkler system, a TI project on the first floor is also impairing the sprinkler system, and another contractor is performing welding in this zone, if a fire started, it could be devastating. You can see why these work activities need to be reviewed and coordinated in a safe manner.

Due to this shift in building condition and livened fire protection systems, the construction fire safety plan must be updated at this transition. The updated construction fire safety plan will reflect the building's operational state, provide new FPPMs and detailed impairment procedures.

Maintenance agreements must be put into place for all fire protection systems in conjunction with the building's property management team. Critical fire protection systems must be maintained at all times without a lapse. Therefore, fire alarm service companies, fire sprinkler service companies, smoke control special inspectors and fire extinguisher service companies must be engaged leading up to substantial completion so the engaged companies can take control without any gap.

19.2 "Small" Building Projects with Big Impacts

Formal projects such as new construction or significant renovations receive a substantial amount of focus on life safety and fire prevention. This is due to the development of detailed drawings and code compliance through the involvement of licensed design professionals and review by AHJs. However, there are many "small" changes that may not go through those channels. What may be deemed "small" or insubstantial could pose fire and life safety risks if not managed properly. Small changes over time, such as the replacement of carpet, chairs, desks and wall coverings, may seem insignificant. Changing out a door, reversing door swings and changing layouts of spaces by furniture or partitions may seem minor. Installation of ceiling decorative elements or other items hung from the ceiling may seem marginal. The storage of items in corridors and exit enclosures may seem like an efficient use of otherwise unused space. Out of negligence, these small changes to the building could pose negative effects that could lessen the life safety level in the building. These changes are also cumulative, so when several modifications are made over several years, the impact can be huge.

Each change in the building must be analyzed closely to ensure there will be no negative impact on the overall life safety level in the building. The question, "does this have any impact to fire protection or life safety?" must be asked with each change.

These changes, upgrades and modifications are not made maliciously, but they could end very poorly. A commercial tenant or property manager may have security concerns and could propose adding a lock or access control system to a particular door. If this door is along the building's egress route, this could be a major problem. These changes would likely be made directly with a vendor or contractor without review from fire protection engineers, code specialists or building department reviews.

Floor, wall and ceiling coverings must meet a minimum flame and smoke spread index based on the occupancy and presence of fire protection systems. If a lesser product is used, the carpet may be more flammable than what is otherwise permitted in the occupancy. This could produce quicker fire and smoke development, significantly reducing the available amount of time for safe exiting for occupants.

Placement of workstations or furniture must be made in a manner that does not negatively impact the building's means of egress systems. New or replacement elements must maintain code compliance. Common issues may include blocking egress routes or exits. The creation of dead-end corridors or exceeding the maximum exit travel distance are also some common results of poorly planned modifications.

19.2.1 Station Nightclub Fire

The Station Nightclub fire was a high-profile fire that occurred on February 20, 2003, at the Station in Warwick, Rhode Island. It was one of the worst fires in modern-day history, with 100 deaths and over 200 injuries. Many couldn't comprehend that a fire of this magnitude could happen in an era of many strict building and fire codes. The night of the fire, Great White, a 1980s heavy rock band, was performing. At the start of their performance, pyrotechnics were ignited as the band started their show. The pyrotechnics quickly ignited the acoustic foam covering on the nearby wall and the fire quickly grew. In only a matter of a couple of minutes, the entirety of the venue was filled with thick, black, toxic smoke. This significantly hindered occupants' vision and breathing.

Most occupants headed straight to the main entrance, the one location they were familiar with, where they entered. This created a stampede and blockage situation, not allowing anyone through. Many were left trapped inside a burning inferno with toxic smoke. The devastating fire left many people asking how and why this happened.

Several investigations into the fire and forensic evidence shed light on the situation. The National Institute of Standards and Technology (NIST) conducted an extensive review and investigation of the fire. Some of the major contributors to the fire's quick growth and impact on the occupants included the following elements[1]:

1 Flammable Foam – the club's walls and ceilings were covered with polyurethane foam for acoustical purposes. There had been several noise complaints from nearby neighbors requiring the owners to address the situation or the city would shut down the business. The foam was highly flammable and did not meet building and fire code requirements for flame and smoke spread characteristics.
2 Pyrotechnics – the use of pyrotechnics within the building as part of the show was not properly authorized. The fire department and the building owner did not give permission for the use, although it would be highly debated whether the owner did or not. The pyrotechnics sprayed hot sparks directly against the walls and ceilings.
3 Exit Door – an exit door that was located adjacent to the stage area that led directly to the exterior was replaced with a door that swung inward, against the direction of egress travel. This door was cited by the fire department several times. There was a sign on the door that read "Door remains closed at all times." There were reports from survivors that claimed a bouncer located at this door said that the door could not be used, and that it was only for the band. This door quickly became engulfed in flames due to its proximity to the fire origin.
4 Occupant Load Exceeded – the club's legal maximum occupant load was debated, as there was contradicting documentation stating 258 or 404 people. In any case, it was determined that these numbers were likely exceeded, as there could have been more than 450 people in the building at the time of the fire.
5 Lack of Sprinkler Protection – the club did not have a fire sprinkler system, only a fire alarm system. A fire sprinkler system was not required at the time, based on adopted codes. The lack of sprinkler protection allowed the fire to develop rapidly, reaching an uncontrollable level before the fire department's arrival.

The installation of acoustical foam on walls, ceilings and doors was done for a good reason: to reduce sound transmission outside of the club to appease neighbors and city officials. However, this good deed would prove to be deadly when the foam that was used was extremely flammable and was one of the main contributors to the fire's quick growth. The foam specifications were not reviewed by engineers, the installation was not reviewed by building and fire department officials; rather, it was done in a vacuum with direct contact with the supplier.

The exit door by the stage was replaced with a door that swung inward. The replacement of this door was not done under the supervision of an engineer, nor was it reviewed by the building and fire department. It was flagged by the fire department in a citation; however, this was after it was already installed. The signage on the door indicated that the door was to "remain closed at all times" and the bouncer was either given direction or took it upon himself to prohibit occupants from using this door.

The tracking of the number of occupants within the building is extremely important in buildings with high occupancies. When occupant loads are exceeded, it puts a strain on the means of egress systems, preventing proper flow of people to exits.

The unauthorized use of pyrotechnics created an ignition source that almost instantly engulfed the foam in flames. The use of pyrotechnics must comply with fire codes, including gaining permission from owners or their representatives, as well as the fire department.

While several factors led to the devastating fire, there seems to be a common theme. Several modifications were made to the building that negatively impacted the life safety level. While it could have been done out of negligence or out of cost-saving efforts, it resulted in leaving the building much more vulnerable than it otherwise would've been. This culmination of defects created a perfect storm of events that led to dangerous, fast-growing fire and smoke development, without fire protection systems to suppress and without adequate means of egress to allow for safe evacuation (Figure 19.1).

Figure 19.1 Station Nightclub Memorial. Photograph by Webster, Joe. November 13, 2003. Wikimedia Commons. Public domain.

19.3 Temporary Events

Temporary uses and events also pose several life safety concerns. Some events may just be a one-day or one-night event. Others may be used for several days or weeks, or sometimes months. Just because these events are deemed "temporary" does not mean there shouldn't be a focus on life safety and fire prevention. In fact, in many cases, these events present more risk than standard building fire prevention. These events may be carried out without or with minimal analysis of means of egress, occupant loading, combustible materials, ignition sources and other hazards.

Once again, it must be asked, "does this have any negative impact to fire protection or life safety?" These details must be closely evaluated to ensure the safety of people, buildings and first responders.

Imagine a company would like to use an area of sidewalk and public space, such as a courtyard, for an activation. A large tent is erected, with tables and chairs and a stage for live music. Propane heaters are provided throughout the space, and sides are added to the tent due to inclement weather, leaving only a single entrance.

This all sounds mostly harmless; however, you cannot think like that if a high level of life safety is desired. Each element must be evaluated to ensure it maintains the level of life safety expected.

Here are several factors to consider for temporary arrangements:

Location – it is important that temporary tents and structures are set up in suitable locations. The placement of the structure in proximity to a building should be evaluated to ensure the structure does not pose an exposure risk to the main building. Creating a separation of 25-50 feet may be appropriate based on the proposed structure. Furthermore, it is important to ensure the structure does not impede any means of egress, exits, fire protection components such as fire department connections, fire pump rooms and other similar life safety-related elements.

Combustible Materials – if there is ever an opportunity to substitute combustible materials for noncombustible or fire-retardant materials, this should be done. Tents and furniture meeting flammability criteria should be utilized.

Ignition Sources – common ignition sources may be heaters, either propane or electric. Candles, firepits and other open flames are also common. Smoking could also present another ignition source. Temporarily running extension cords could also become overloaded and become a fire risk. These items must be carefully evaluated and located to ensure they do not become a fire hazard.

Means of Egress – temporary events must always be evaluated for occupant load capacity and exiting capacity to ensure the exit arrangement is sufficient for the event and the number of people. This is particularly important at large events where there could be 500–1000 or more people.

Temporary does not mean any leniency on evaluation for fire protection and life safety. I often spend a significant amount of time analyzing these types of events since there is often minimal information or plans with short review durations.

19.3.1 *Tomorrowland Music Festival Fire – Belgium*

A stark reminder of the serious fire safety risks in temporary events was seen at the Tomorrowland music festival in Belgium in July 2025. Just two days before the start of the highly attended festival that spans multiple weekends and brings in more than 400,000 people, a fire destroyed the main stage. The main stage was a robust structure for the performers, but it also consisted of themed decorations and backdrops that made the structure very large. Mountains, sculptures and castles soared from the stage. The cause of the fire remains under investigation. The rapid growth of the fire and overall magnitude of the fire were greatly influenced by the construction materials of the stage and decorations. Polystyrene and wood elements were heavily used to construct many of the stage elements.

As devastating as the fire was, the only silver lining was the fact that the fire occurred before thousands of people attended the event. If the fire had transpired during the event, the impacts could have been a lot more detrimental. The event had to pivot and quickly reconstruct another main stage to allow the event to carry on.

This fire highlights the importance of constructing temporary structures with fire safety considerations. It is the responsibility of the event organizer to ensure an acceptable level of life safety, especially at high-occupancy events such as music festivals. The development of plans, permitting processes, and evaluation of construction materials may have been abbreviated due to the temporary nature of the event. While the event may be temporary, the impacts of a fire on occupants could lead to very permanent results. Therefore, fire prevention must be considered early, incorporated into the design and carried out.

Reference

1. Grosshandler, William, Nelson Bryner, Daniel Madrzykowski, and Kenneth Kuntz, *Report of the Technical Investigation of The Station Nightclub Fire (NIST NCSTAR 2), Volume 1*, National Construction Safety Team Act Reports (NIST NCSTAR), National Institute of Standards and Technology, June 30, 2005. https://www.nist.gov/publications/report-technical-investigation-station-nightclub-fire-nist-ncstar-2-volume-1

Chapter 20

Incident Management and Post-Incident Evaluation

20.1 Managing Fire Events

The role of the Fire Protection Program Manager (FPPM) during a fire emergency becomes extremely crucial to ensure a rapid response and to orchestrate the incident management protocol. The FPPM's goals include ensuring the safety of occupants, minimizing property damage and limiting harm to first responders. Ensuring there are clearly defined roles and responsibilities, a communication framework and evacuation procedures is essential. A common theme for successfully carrying out an incident management plan is rapid communication.

The sooner that the fire is identified, and the faster the report of the fire to the FPPM or supervisor, and the earlier the fire is reported to the fire department or other first responders, the higher the chances of containing the fire and minimizing the impact of the fire on occupants and the building. There must be communication channels established between workers and their supervisors, between the supervisors and the FPPM, and between the FPPM and first responders.

When a fire is detected, it must be reported to the supervisor, who will report it to the FPPM. The FPPM must call 9-1-1 and report the emergency if not reported by the worker or supervisor. Now, while all other response activities are carried out, the fire department is in motion responding to the building. This is critical since often so much precious time is wasted trying to get different individuals' eyes on the fire to extinguish it without the help of authorities. During this time period, the fire grows from a small fire to a fire that is much harder to control.

Next, in tandem with first responders on their way, is to notify building occupants of the fire emergency to begin the evacuation. In buildings with fire alarm systems, activation of a pull station will serve this purpose. In buildings without fire alarm systems, the notification procedures found in the construction fire safety plan should be used. This may include a temporary fire alarm system with alarms that can be activated. In other cases, it could simply be air horns in addition to verbal communication to workers. Designating individuals as incident managers will ensure essential roles are carried out during an event. Having individuals notify occupants of an emergency and direct them to exits will expedite the evacuation process.

DOI: 10.1201/9781032720210-23

Simultaneously with the evacuation process, if safe to do so, an attempt can be made to suppress the fire with a fire extinguisher by those who are trained and qualified. If the fire is rapidly growing, emitting toxic smoke or if other dangers exist, all occupants should evacuate and leave it to the trained professionals of the fire department to fight the fire.

It is important that the FPPM gathers as much information as possible to provide both to the dispatch and to the fire department as they arrive on scene. It sounds like common sense, but a really important piece of information is the building address. Buildings under construction may have a different permitting address or a range of addresses, or many different entrances or gates to the jobsite. Having explicit directions and address information will improve the response time. Information on the location and other details of the fire is equally important. Providing the floor level, location and providing access will help first responders as they make their initial search into the building.

Information on active fire protection systems is also very important. The firefighting response approach may change if there are active standpipe systems and fire sprinkler systems. Communicating the location of fire department connections, standpipe hose connections and active and inactive systems will provide the fire department with useful information.

The FPPM must work with the fire department to properly manage building infrastructure. Electrical, HVAC, gas and other systems may be requested to be shut off to either protect them or to prevent them from exacerbating the fire. The FPPM must know where these rooms exist, how to gain access, and how to operate the systems to deactivate them.

The FPPM must work with supervisors to ensure all occupants representing all companies, trades and subcontractors are accounted for. This is particularly important in large buildings where hundreds of occupants from different companies occupy.

The rest of the response is left to the fire department. The FPPM should keep themselves available to the fire department for questions, building information, and other contacts.

20.2 Post-Incident Evaluation

After the fire is extinguished, smoke has settled and other associated risks have been alleviated, the next phase is the post-incident evaluation. An initial review with key stakeholders such as the fire department, building management, general contractor, FPPM, and relevant subcontractors should take place. All current information should be presented and gathered, as different stakeholders may hold different information and perspectives. Potential causes should be discussed. The performance of fire protection systems must be evaluated. A close look at the detection and notification of the fire is important.

After the preliminary evaluation, a continued investigation into the fire should be conducted. Hearing from witnesses and others involved can help piece together the events that occurred. The more information and timeline of events that are understood, the better the postmortem analysis can help to detect failures.

The findings from the post-incident evaluation should be carefully analyzed so fire prevention protocols can be modified or improved. Corrective actions should be taken to address any failures in the fire prevention plan. Fire protection systems should be examined to see if they operated as designed. If not, it should be explored where the failures were. The communication framework should be similarly reviewed to see if there were any breakdowns in these channels as well.

In the aftermath of a fire, another important aspect is restoring building systems quickly and safely. If the fire alarm system was activated, it should be reset and evaluated for any damage or defects. The fire sprinkler system, if it were activated, must follow a few steps to have it become operational again. It is possible that the entire system or a portion of the system was zoned off by a control valve to stop the flow of water once the fire was under control. The activated and fire-exposed sprinkler heads and piping must be analyzed and replaced as needed so the system can be reactivated and filled with water. The delay in getting the system operational again presents an unnecessary risk in an already fire-damaged building. Other fire protection systems must follow a similar evaluation to ensure they are fully restored. Other building systems that were impacted, such as electrical, HVAC, and gas systems, must be restored by licensed workers and with the coordination of utility companies.

Collaboration with insurance companies becomes important during the aftermath as well. The insurance company may want to carry out its own investigation or inspection. Therefore, although the initial reaction may be to remediate and restore the building, it must be in accordance with the insurance company's protocols.

The emotional and psychological impacts a fire has on victims, employees, tenants or others that experience the negative effects of the fire must not be forgotten. This is extremely important in cases where there may have been loss of life or injuries. However, it is important in all instances to consider the trauma that can be caused by a fire. All those impacted should be provided with access to counseling services to help cope with the long-term effects of a fire.

Chapter 21

Home Considerations

21.1 Home Fire Safety

According to the National Fire Protection Association's (NFPA) November 2024 Report on Fire Loss in the United States During 2023, the most injuries and casualties from fire are at residential homes.[1] Therefore, home fire safety must be prioritized. Significant fire protection features are provided in commercial buildings; however, residential homes are often overlooked, even though each of us is likely to face our biggest threat from fire while in our homes. We spend significant time at home, including cooking and eating, working from home, sleeping and otherwise occupying our homes.

While some considerations for fire safety are made when the house is originally constructed, very few are properly maintained. Most houses are required to have smoke alarms, fire extinguishers, and fire-rated elements for garage or dwelling separation. If these features are not understood by homeowners, it is unlikely they are adequately maintained. If not maintained, they will not provide the protection needed and will only provide a false sense of security.

When construction or other home improvement projects are taking place in our homes or within occupied residential buildings, we must take a similar approach to preventing fire as is done with larger-scale commercial projects. Construction activities or other common household hazards must be understood to mitigate the risk of a fire.

21.2 Common Fire Protection Features Found in Residential Homes

Smoke Alarms – smoke alarms must be provided on every level of the home and in close proximity to the bedrooms. Smoke alarms must be regularly tested; most manufacturers and codes recommend once a month. Smoke alarms typically have a lifespan of 10 years. Therefore, the date of manufacture must be located on each device and replaced at no longer than 10-year intervals.

Construction considerations – when home projects take place either by the homeowner themselves or by a hired contractor, some work may produce dust

DOI: 10.1201/9781032720210-24

or other small particles that could interfere with the smoke alarms. Examples include sanding floors, walls and furniture. Additional activities include spraying paint, and cutting wood, tile and other materials. Therefore, it is often encountered in most home projects. Consideration should be given to managing smoke alarm impairments. Smoke alarms should be protected with covers during this dust-generating work period. Sometimes the smoke alarm manufacturer includes a protective cover when the device was originally purchased for this purpose. If not, covers should be able to be found at home improvement stores or online.

The most important part is that the smoke alarm covers are removed immediately after the completion of work each day. It is critical that the smoke alarms remain functional at all times when dust-generating work is not occurring. If smoke detectors remain covered for extended periods of time, extending multiple days, weeks or even months in some projects, occupants residing in the house are lacking the most crucial fire notification system that their lives depend on.

Carbon Monoxide Alarms – carbon monoxide alarms must also be provided on each level of the home. These should be replaced in accordance with the manufacturer's recommendations; however, it is best practice to replace them every 5 years. The alarms must typically be installed in close proximity to bedrooms, which is informed by the applicable building and fire codes.

Construction considerations – carbon monoxide alarms must remain active at all times, even during construction activities. Additional considerations must be made if there is an increased risk of carbon monoxide due to the construction equipment that is used. If generators, pressure-washers or other fossil fuel-burning equipment are used, they should be placed outside the house with unobstructed open-air surrounding it.

General considerations – during power outages in the wintertime, a temporary heating solution is sometimes needed. However, introducing a portable fossil-fuel burning heater into the home can pose a significant risk of carbon monoxide poisoning as well as other fire hazards. Heaters must only be used in fully ventilated spaces where carbon monoxide can be monitored, detected, and alerted to occupants.

Fire Extinguishers – all homes must be provided with appropriately specified fire extinguishers. Often, a multipurpose fire extinguisher (A, B & C) makes sense in most applications. Extinguishers should be provided in the kitchen and other cooking locations, the garage and at each level of the house. Fire extinguishers must be inspected regularly to ensure they are still in good working order. Homeowners should ensure that the pressure is still adequate. This can typically be achieved by reading the pressure gauge and ensuring the pressure is within the acceptable range. Most present-day fire extinguishers are provided with a gauge with a delineated green-acceptable range and a red-unacceptable range to easily determine the status. Access to fire extinguishers should be planned to ensure they are located where the hazards exist and where they can be accessed without obstructions.

Construction considerations – fire extinguishers must be provided in all areas under construction in the house. There is an increased risk of fire in these areas due to combustible materials, electrical system installations, soldering of plumbing systems or temporary heat.

Fire-rated Elements – in multi-dwelling houses (townhouses, duplexes, apartment buildings, condominiums and similar), there is likely a fire-rated separation between dwelling units. When houses are provided with garages, similar fire-rated separation is provided to protect the home from the increased hazards found in a garage. Typically, these rated assemblies are provided by way of rated walls, rated doors and firestopped penetrations. It is important that homeowners understand these fire protection features so future modifications to the home do not jeopardize these passive fire protection systems.

Construction considerations – Homeowners may need to modify fire-rated elements for various reasons; therefore, it is important that they recognize these areas and consider them when making decisions. For example, if the garage door between the interior of the house and the garage needs to be replaced, repaired or otherwise modified, the required fire-rating of this door must be considered and maintained. If a standard, non-fire-rated door is installed in this location by negligence, this will create a weak point in the fire-rated separation between the house and the garage and will not maintain the 1-hour or specified fire rating. Other examples include penetrations through the garage wall and into the house. If these are not properly firestopped and sealed, this will provide a pathway for fire and smoke to travel into the house much faster, resulting in a faster-developing fire.

Means of Egress – houses are typically provided with multiple exits to provide occupants with more than one escape route in an emergency situation. These include front doors, side doors, basement bulkheads, slider doors, emergency escape windows, and other doors that provide access to the exterior.

Construction considerations – all means of egress serving the house must be maintained during construction projects. If the proposed work will obstruct an exit, additional considerations and mitigation measures must be made. For example, if a deck is being replaced where there is an exit leading onto it, there will be a period of time during which the exit will be unusable. The shutdown should be minimized by prioritizing the completion of the portion of the deck at the exit.

General considerations – Exit doors must always remain unlocked from the interior of the house. Doors that are locked or fastened shut are extremely dangerous since they eliminate one of the critical escape routes. Furthermore, baby and pet gates must not obstruct exiting routes that will slow down evacuation times in an emergency.

21.3 Common Fire Hazards

Cooking – cooking remains the number one cause of home fires according to the National Fire Protection Association's (NFPA) November 2024 Report on Fire Loss in the United States During 2023[1]. It is important that kitchens are closely monitored when cooking is taking place. Unattended ovens, stoves, and other kitchen cooking appliances can start fires that could otherwise be suppressed in their infancy by water or fire extinguishers.

Grills, smokers and similar items must be located away from the house and combustible items. They should be only used in areas with adequate ventilation. These items must also be closely attended to even after their use while they cool down.

Smoking – smoking in homes is another major fire contributor. Smoking should be limited to outdoor spaces only. An established safe discard location must be developed. Smoking indoors presents many hazards due to the smoldering characteristics of cigarettes and similar paraphernalia. These items can remain an ignition source well after these items have been discarded.

Fireplaces and Fire Pits – indoor fireplaces must be carefully used to ensure they do not present a fire risk to the rest of the house. When in use, fireplaces must remain monitored to ensure the fire does not grow out of control or sparks do not escape the fireplace. Combustible items on mantles, hearths or in close proximity to the fireplace must be relocated when the fireplace is in use. Fireplaces and chimneys must be regularly inspected and cleaned to ensure they remain in good condition.

Outdoor fire pits must be located away from the house and other structures. Often, towns/cities have clearance requirements and sometimes require permits to be pulled for their use. Similarly, these fire pits must be attended at all times and the fire must be extinguished before they are left.

Electrical Equipment – electrical equipment, wires, and plugs present a fire hazard if used incorrectly. Clearances must be maintained at all electrical panels and other equipment. Wires should not be exposed and disconnected equipment must have its electrical connections safely terminated without leaving loose wires. Receptacle outlets must not be overloaded with multiple 3-way adapters or extension cords.

Home Heating – portable electric space heaters are common in colder climates in the wintertime. These must be used in accordance with the manufacturer's recommendations, including providing adequate clearance around them and the power supplies. Although modern space heaters are provided with safety mechanisms to prevent overheating and fires, they should always be unplugged when not in use to prevent any malfunction that could potentially cause a fire. If some areas of the home are consistently colder than others, it should be addressed by engaging an HVAC contractor to provide a more permanent solution rather than depending on space heaters on a long-term basis.

Candles – candles can be dangerous for obvious reasons. You are creating a small flame that burns for long periods of time within the house. LED candles, diffusers or other elements are safer alternatives to candles. However, when traditional candles are used, they should be used only where they can be monitored and where they are on a level surface, away from any furnishings.

21.4 Home Fire Safety Checklist

A home fire safety checklist, similar to the one below, should be utilized by residents to ensure the fire protection features that are provided within the home are maintained adequately (Table 21.1):

Table 21.1 Home Fire Safety Checklist

Home Fire Safety Checklist
Fire Protection Features
Smoke Alarms
• In good working order (connected to base, uncovered, unobstructed, batteries ok). • Smoke alarms < 10 years old. • Tested monthly.
Carbon Monoxide Alarms
• In good working order (connected to base, uncovered, unobstructed, batteries ok). • CO Alarms < 5 years old. • Tested monthly.
Fire Extinguishers

• Sufficiently located at:
  • Each level of home
  • Kitchen
  • Garage
• Inspected monthly
  • Pressure maintained.
  • Access unobstructed.

Fire-rated Elements
• The garage swinging door automatically closes and latches. • No visual compromises observed made to fire-rated assembly.
Means of Egress
• Exits remain unobstructed and unlocked in the direction of egress. • Exits maintained at the exterior to prevent snow/ice and other elements from inhibiting use of exits.

(Continued)

Table 21.1 (Continued)

Home Fire Safety Checklist
Household Hazards Cooking • Cooking operations monitored at all times. • Grills/smokers and other outdoor cooking appliances to be located away from house and combustible items and in well-ventilated, open-air areas. Smoking • Smoking to be limited to outdoor areas only. • Discarded smoking materials to be placed in appropriate canisters, located away from house. Fireplaces • Must remain monitored at all times until fully extinguished. • Combustible items and furnishings must be removed from mantles, hearths and surrounding areas. Outdoor Fire Pits • Must remain monitored at all times until fully extinguished. • Placed away from house and other structures. • Permits obtained as applicable. Electrical • Electrical panels and other equipment must be maintained with proper clearance surrounding it. • Receptacle outlets must not be overloaded with 3-way plugs or extension cords. Heating • Space heaters are located on flat, level surfaces. • Adequate clearance provided around heater to furnishings and other combustible materials. • Heaters remain unplugged when not in use. • When in use, heaters only in areas where occupied and monitored. Candles • Candles located on flat, level surfaces. • Adequate clearance provided around candle to furnishings and other combustible materials. • When in use, candles only in areas where occupied and monitored.

Reference

1. Hall, Shelby. *National Fire Protection Association (NFPA) Report on Fire Loss in the United States*. November 2024. https://www.nfpa.org/education-and-research/research/nfpa-research/fire-statistical-reports/fire-loss-in-the-united-states.

Chapter 22

Conclusion

22.1 Final Thoughts

I believe all people who are involved in the design, construction and management of buildings must have a fundamental understanding of fire prevention. It must be something that is thought of early, reevaluated often, and adapted continuously. When there is buy-in from the entire project team, a high level of life safety can be achieved.

The Fire Protection Program Manager (FPPM) has the critical role of carrying out the construction fire safety program and planning and maintaining fire protection systems. The FPPM must continue to learn and adapt as construction techniques, methods, materials and technologies evolve. The FPPM must stay ahead of the curve to anticipate the risks and provide mitigation plans. Just as the role has evolved in the past several decades, it will continue to progress in a similar fashion.

I believe the capabilities of fire protection systems in buildings will continue to become more effective, more reliable and more robust. The FPPM must familiarize themselves with the latest trends and technologies and design best practices.

While the role is sometimes underappreciated, unrecognized, or taken for granted, that does not change its importance. Just like the world of construction safety, construction fire safety is also very hard to quantify the benefits, as it is hard to evaluate things that did not happen. However, when you look at both the frequency and impact of construction fires occurring in the news, construction fire safety is not going anywhere; in fact, it will be more relevant as the years go on.

I believe everyone should consider life safety and fire prevention with each decision that is made on their building. Once we see all the direct and indirect impacts these decisions have on critical life safety systems and their impact on the well-being of occupants, then we will start to see a change. Historically, it unfortunately takes devastating fires, loss of life and property loss to make changes. Those who can do the right thing now will stay ahead of the risks associated with fire in buildings and will benefit long-term from it.

Embrace the silent success of fire prevention – not measured by what was lost, but by what was avoided.

DOI: 10.1201/9781032720210-25

Part IV

Appendices

Supplemental Information

Appendix A

Boston Seaport Case Study

Appendix A.1 A Major Development in Boston's Waterfront

The Seaport neighborhood in Boston, Massachusetts, has seen major development in the past decade to become one of the most desired areas of Boston. Some of the most successful tech companies, life science companies, restaurants and retailers have landed in this neighborhood. The Seaport has found success by constructing a well-balanced neighborhood with multiple uses, and enhanced by abundant green spaces and parks.

Buildings in the Seaport neighborhood are of similar height, about 250 feet, due to FAA regulations, due to the proximity to the Boston Logan Airport across the harbor in East Boston. The number of stories varies based on floor-to-floor heights; however, most buildings are between 16 and 20 stories. Most buildings contain sub-grade parking garages, mixed retail on the lower podium levels, and office, life science or residential uses occupying most of the upper levels.

We are going to explore a specific project within the Seaport development: One Seaport Square. This project involves two high-rise towers constructed on top of a common, three-level, sub-grade parking garage. The towers are referred to as Block B and Block C; however, the building was constructed as a single building. Block B consists of three levels of mixed-use, multi-tenant retail, spanning levels 1, 2 and 3. Residential apartments and amenity space occupy levels 4 through 20. Block C consists of three levels of mixed-use, multi-tenant retail, spanning levels 1, 2 and 3. Residential apartments and amenity space occupy levels 4 through 22.

The unique part about this project is the multiple owners, general contractors and engineers who cohesively built this building. The residential portion of the building and a portion of the parking garage are owned and operated by a residential developer. Whereas the three levels of retail in each tower spanning across almost 250,000 SF is owned and operated by another retail, commercial and life science-centric developer. There are approximately 25 different retail tenants spanning the building, ranging from restaurants to a movie theater, nightclubs, a barber, fitness and several dry retail stores.

To complicate matters further, the building underwent a phased occupancy to open specific areas of the building first while construction continued elsewhere. As base building construction was nearing completion, there was a shift in focus to retail construction. With several leases signed and tenants eager to begin their build-outs to jump on the success of the Seaport neighborhood, retail construction started. Many retail tenants engaged their own general contractor based on their preference or the result of a competitively bid project. This resulted in many different general contractors working in the building, alongside each other, to complete these various buildouts. These were extensive buildouts too, often involving vertical transportation, modifications to the building structure, modifications to fire protection systems, installation of kitchen equipment, acoustic floor and ceiling systems on top of the typical MEP and finishes that come with a tenant improvement (TI) project. Meanwhile, the base building general contractor was still onsite while retail construction was underway. Phased areas of the building were still under major construction, while some areas were going through punch list and commission efforts.

The fire and building departments, as well as the project team, knew that with all of this construction happening simultaneously, it had to be managed carefully with construction fire safety as a top priority. This became especially important as some areas of the building became occupied by residents now living there, with significant construction occurring elsewhere. That is where I entered the scene.

I was engaged by the project team to serve as the owner's Fire Protection Program Manager (FPPM). I was tasked with developing and implementing the master construction fire safety program across the base building construction and retail construction. This included interfacing with the fire department, managing impairments and hot work, conducting site inspections, and managing tie-ins and modifications to the building's critical life safety systems.

My first several weeks on the job in January 2017 were spent familiarizing myself with the building, meeting key project team members and developing a construction fire safety program. I knew, just like any other operational program that involves people, the success of the program would depend on the attitude the project team has towards it. Over the course of the first few months on the project, a comprehensive construction fire safety plan was developed. It included processes for conducting fire protection impairments, hot work activities, waste management, inspections, pre-fire plans and other critical elements. This master plan would become the foundation for project-specific construction fire safety plans developed by each retail tenant's team. Each project-specific plan was reviewed to ensure alignment with the overall program.

As part of the onboarding, orientation to the project, each retail tenant was required to submit a project-specific construction fire safety plan. This plan would dive into the specifics of the project. For example, if the project involved a second-floor restaurant, the plan would include construction mitigation plans for the welding and impairments required to accommodate a new stair and elevator. It would also consider the means of egress serving the tenant space. Site access, command posts were all considered based on the intricacies of the project. The project-specific plans also designated a project-level FPPM and an alternate. These FPPMs became my counterpart in managing this program in the building. Project-level construction fire safety plans often took several revisions until they met the expectations and aligned with the overall program. Chapter 4 provides further information regarding the required content in master and project-specific construction fire safety plans.

Before the commencement of each retail construction project, a dedicated onsite meeting was conducted with the project level FPPM to review critical documents, plans, procedures and to locate valves, terminal cabinets associated with fire protection systems. These initial meetings set the project off on the right foot and demonstrated to the project team the priority given to construction fire safety. This is also a good opportunity to review city/town-specific requirements relating to fire prevention, such as hot work, smoke detection, bagging permits and other permitting requirements. This was especially important when dealing with out-of-state contractors who may be accustomed to different city guidelines. I have found that this initial meeting makes the rest of the project go smoothly and opens the door for strong communication, which is necessary for the success of the program.

Recurring FPPM meetings were established to review the construction fire safety program. Key items reviewed included impairments, fire alarm tie-in and testing needs, hot work activities and other observations from the week's work. These meetings proved to be extremely valuable. An FPPM representing each project attended and would review their project's status. The projects were all under rigorous schedules to meet grand openings, holidays, and other critical dates. Impairments to fire protection systems became critical for all the tenants, as it would allow them to modify fire protection systems to continue with their build-outs. Conflicting impairment requests would need to be reviewed and awarded in a fair manner. Chapter 17 provides additional information on FPPM meetings, sample agendas, and other helpful resources.

The final stages of the retail projects involved fire alarm tie-ins and programming to the base building system. With the building now occupiable, fire alarm testing quickly became a nuisance to residents and occupants. Each tenant underwent the initial fire alarm tie-in, programming and pre-test. If that was successful, a fire department fire alarm acceptance test was scheduled. That meant that for every retail tenant, at least two fire alarm tests needed to occur if all went well. That created almost daily fire alarm testing. Fire alarm testing notifications were developed to properly notify impacted areas of the building. The only thing worse for a retail tenant's business operations and residents than fire alarm testing is unexpected fire alarm testing without notification and preparation. Fire alarm tests were consolidated together to minimize building impacts and bypasses were implemented to minimize the area of audible and visual alarms. Chapter 6 takes a deeper dive into fire alarm system planning and integration with retail tenants.

As each retail tenant sought to complete construction and obtain a certificate of occupancy (CofO), a number of base building testing and inspection documents were needed. Testing and inspection reports, such as fire alarm, sprinkler, fire pump, generator, smoke control are all needed and must be up-to-date without any deficiencies noted. On top of ensuring that this ongoing maintenance is occurring, it is also just as important to ensure the testing and inspection reports are organized in a way that can make sharing the reports easy. Part II expands on these testing and inspection requirements.

Appendix B

Construction Fire Safety Plan Outline

Appendix B.1 Introduction

This chapter provides a templated outline for a construction fire safety plan that can easily be adapted to any project. It is important that all hazards and fire protection systems are considered. If some contents of the plan are not applicable, the section should still exist and identify that it does not apply to the project. This approach ensures items are not overlooked; rather, they are considered and deemed not applicable. In the template below, you will see recommended language for applicable and non-applicable items.

[Name of project]
[Address of project]
[Associated building and fire permit numbers]
Construction Fire Safety Plan

[Image of Building or Rendering]

	Name	*Company*	*Email*	*Phone*
Fire Protection Program Manager				
Alternate Fire Protection Program Manager				

Table B.1 Impairment Log

Date	*Time out*	*Time in*	*Level*	*Tenant*	*Contact Information*	*Contractor*	*Impairment Type*

Table B.2 Hot Work Log

Date	*Time*	*General Contractor/ Tenant*	*Subcontractor*	*Type(s) of Hot Work*	*Location(s)*	*Name of Fire Watch*	

Table of Contents:

1. Applicable Codes and Standards:

Code/Standard	*Applicable Code and Standard and Edition*
Building Code	• [Edition, Standard/Code, and any amendments applied]
Fire Code	• [Edition, Standard/Code, and any amendments applied]
Construction Fire Safety Codes	• [Edition, Standard/Code, and any amendments applied]
Fire protection and Life Safety System Codes	• [Edition, Standard/Code, and any amendments applied]
Electrical Code	• [Edition, Standard/Code, and any amendments applied]
Testing and Inspection Standards	• [Edition, Standard/Code, and any amendments applied]

2. Roles and Responsibilities of Stakeholders:

* *Example of Owner's Master Construction Fire Safety Plan:*

Ownership:

[The owner has developed this plan to serve as the master construction fire safety plan for the building, {insert address}. The owner has designated {insert name and company} to serve as the owner's Fire Protection Program Manager and {insert name and company} to serve as the alternate Fire Protection Program Manager. The Fire Protection Program Manager (FPPM) and alternate will implement the contents of the construction fire safety plan and coordinate with the project-level FPPMs to ensure all ongoing construction activities meet the plan.]

General Contractor:

[The general contractor will abide by the requirements of the master construction fire safety plan. The general contractor will also create a project-specific construction fire safety plan. This project-level plan will identify a project-level FPPM and alternate that is responsible for carrying out the contents of the master and project-specific construction fire safety plan. The project level FPPM must coordinate and communicate as necessary to carry out the work in a safe manner.]

Subcontractor:

[All subcontractors and workers on the project must abide by the requirements of this master construction fire safety plan. It is the responsibility of the general contractor to ensure the requirements are effectively communicated to all workers on-site. Any deviations from the master construction fire safety plan must be approved by the Owner's FPPM.]

**Example of Project-Specific Construction Fire Safety Plan*

Ownership:

[The owner has designated {insert name and company} to serve as the owner's Fire Protection Program Manager and {insert name and company} to serve as the alternate Fire Protection Program Manager. The owner has developed a master construction fire safety plan for the building, {insert address}.]

General Contractor:

[The general contractor has developed this plan to serve as the project-specific construction fire safety plan for the project {insert name of project and specific address} within the building, {insert address}. The general contractor has designated {insert name and company} to serve as the project-specific Fire Protection Program Manager and {insert name and company} to serve as the alternate Fire Protection Program Manager. The general contractor acknowledges the owner's master construction fire safety plan. The FPPM and alternate will implement the contents of the master and project-specific construction fire safety plan and coordinate with the owner FPPM to ensure all ongoing construction activities meet the plan.]

Subcontractor:

[All subcontractors and workers on the project must abide by the requirements of this master construction fire safety plan. It is the responsibility of the general contractor to ensure the requirements are effectively communicated to all workers on-site. Any deviations from the master construction fire safety plan must be approved by the Owner's FPPM.]

3. Construction Site Hazards:

Smoking:

[Insert smoking policies on construction site. If smoking is prohibited everywhere on site, it should be stated here. If designated smoking areas are implemented, a site plan should be inserted here showing the permitted locations. Protocols for disciplinary action for offenders should be outlined in this section. Controls to prevent smoking should be explained in this section, such as "no smoking" signage, inspections, etc.]

Temporary Heating:

[Temporary heating equipment is not anticipated as part of the project. Temporary heating equipment will be prohibited from the construction site and monitored by the FPPM to ensure compliance with this requirement.]

[Temporary heating equipment is anticipated as part of the construction project to provide climate-controlled areas to allow for construction activities to continue during the cold time periods of the project. A {insert type of heater} will be utilized to heat the building. The heater will be located at {insert location} as shown on the pre-fire plan in the appendix. Clearance around the heater will be maintained in accordance with the manufacturer's recommendations and applicable codes and standards. Heating equipment will be monitored by the FPPM when in use and will not be left unattended.]

A carbon monoxide monitoring plan will be implemented. The monitoring program includes several battery backup CO detectors located throughout the building where the equipment is present and other atmospherically connected areas. In addition, carbon monoxide detectors with parts per million display readings should be utilized to track fluctuations in carbon monoxide concentrations. These concentrations should be recorded in a log. The site will be evacuated if elevated levels of carbon monoxide are recorded.]

Cooking Equipment:

[Cooking equipment is not anticipated as part of the project. Cooking equipment will be prohibited from the construction site and monitored by the FPPM to ensure compliance with this requirement.]

[Cooking equipment is anticipated as part of the construction project. A designated approved cooking location will be established at the {insert location} as shown on the pre-fire plan in the appendix. The designated cooking location will be maintained, including clearance around the cooking equipment in accordance with the manufacturer's recommendations and applicable codes and standards. Only the following types of cooking equipment are permitted on-site:

- [microwave ovens]

The following cooking equipment is prohibited from the site:

- [propane and natural gas grills]
- [charcoal grills and similar cooking equipment]
- [Burners, griddles]

Cooking equipment will be monitored by the FPPM when in use and will not be left unattended.]

Lithium-ion Batteries:

[The use of lithium-ion batteries is not anticipated as part of the project. Lithium-ion batteries will be prohibited from the construction site and monitored by the FPPM to ensure compliance with this requirement.]

[It is anticipated that there will be lithium-ion batteries on site as part of the construction project. Personal transportation vehicles and power tools containing lithium-ion batteries are expected. A designated approved storage location for scooters, bicycles and similar transportation types containing lithium-ion batteries will be established at the {insert location} as shown on the pre-fire plan in the appendix. These designated storage areas will be kept clear of combustible materials and fuel storage. The designated storage location will be maintained, including clearance around the storage area in accordance with the manufacturer's recommendations and applicable codes and standards.

Power tools and other equipment containing lithium-ion batteries will be utilized by subcontractors. Subcontractors and their workers are responsible for the safe management of the lithium-ion batteries. Only listed and approved chargers and batteries will be used. Batteries and tools that become damaged will be removed from use and the jobsite. The FPPM will monitor and enforce the safe management of batteries on the construction site.]

Combustion Engine Equipment:

[Combustion engine equipment is not anticipated as part of the project. Combustion engine equipment will be prohibited from the construction site and monitored by the FPPM to ensure compliance with this requirement.]

[Combustion engine equipment is anticipated as part of the construction project. {Insert equipment type} will be utilized onsite to support construction operations. The equipment will be placed in an area with adequate ventilation with sufficient clearances to combustible materials in accordance with the manufacturer's recommendations and applicable codes and standards. The location of the equipment is shown on the pre-fire plan in the Appendix. Where the equipment is located within the building, a ventilation approach will be implemented, including {natural or mechanical ventilation} to ensure levels of carbon monoxide and nitrogen dioxide are maintained at safe levels.]

4. Fire Protection Systems:
This section outlines fire protection systems provided within the building.

Fire Sprinkler System:

[The existing building is not provided with a fire sprinkler system. A fire sprinkler system is not planned as part of the work.]

[The existing building is provided with a fire sprinkler system throughout. The sprinkler system is comprised of a wet sprinkler system serving {insert locations} and a dry sprinkler system serving {insert locations}. Control valves are provided to isolate portions of the sprinkler system. {insert description of location of various control valves}.{Describe system components, hazard classifications, integration with other fire protection systems, etc.}]

[A fire sprinkler system is planned as part of the construction project. The installation and activation of the system will be prioritized to ensure the system becomes live as soon as practicable.}. {Describe system components, hazard classifications, integration with other fire protection systems, etc.}]

Fire Alarm System:

[The existing building is not provided with a fire alarm system. A fire alarm system is not planned as part of the work.]

[The existing building is provided with a fire alarm system throughout. The fire alarm system is comprised of a {insert model of fire alarm system} fire alarm control panel located at {insert location}. The system includes a combination of initiating and notification appliances. {Describe system components and insert specific devices present at the building, such as smoke detectors, heat detectors and pull stations}.

[A fire alarm system is planned as part of the construction project. The installation and activation of the system will be prioritized to ensure the system becomes live as soon as practicable.}.{Describe system components, proposed devices and integration with other fire protection systems, etc.}]

Fire Extinguishers:

[Fire extinguishers will be provided throughout the work area by the general contractor. Insert requirements for travel distances, fire extinguisher types, minimum sizing, quantities and signage in accordance with applicable codes and standards. Fire extinguishers will be installed as shown on the pre-fire plan in the Appendix. The FPPM is responsible for inspecting the fire extinguishers to ensure they remain in working order and accessible at all times.]

Smoke Control Systems:

[The existing building is not provided with smoke control systems. Smoke control systems are not planned as part of the work.]

[The existing building is provided with smoke control systems. {Insert smoke control systems present in the building, location and how they function.} {Pressurized stairwells and elevators should be listed here if applicable. Atrium smoke control systems and other smoke exhaust systems should be outlined if applicable.

[A smoke control system is planned as part of the construction project. The installation and activation of the system will be prioritized to ensure the system becomes live as soon as practicable.}.{Describe system components, proposed devices and integration with other fire protection systems, etc.}]

5. Fire Protection Impairment Procedures:
[No impairments to any fire protection or life safety system are anticipated as part of the project]

[As part of the construction project, impairments to the existing fire protection systems are necessary to accommodate the new layout. Impairments will be carried out in accordance with this section:] (Table B.1)

Fire Sprinkler Impairments:

[No fire sprinkler system impairments are anticipated as part of the project.]

[It is anticipated that impairments to the fire sprinkler system will be necessary as part of the project. Due to the installation of {insert specific elements such as ceiling assemblies, partitions, etc.}, modifications to the sprinkler system are necessary to maintain adequate sprinkler coverage. {Describe sequencing approaches that address how sprinkler coverage will be maintained throughout the various stages of construction. For example, describe how pendent sprinkler heads located within a drop ceiling will be turned upright during construction while the ceiling is removed. Then returned to a pendant position when the new ceiling is installed.}

{Specific fire sprinkler impairment procedures should be outlined here}. Fire sprinkler impairments must be coordinated with the owner's FPPM by the submission of a fire protection impairment request form. Fire sprinkler impairments must be carried out in accordance with the following protocols:

1 Analysis of work tasks to determine modifications required for the automatic sprinkler system.
2 Impairment request form submitted from contractor performing work to project FPPM.
3 Project FPPM to review requests against other work occurring, extent of impairment and duration.
4 If approved, project FPPM to schedule fire alarm service vendor to disable the requested sprinkler zone for the identified duration.
5 Project FPPM to notify the project team, including property managers, building engineers, superintendents, of scheduled impairment to ensure the greater team is aware.
6 Verification of correct/required devices by the fire alarm service company and contractor to ensure associated flow switches and tamper modules are disabled for sprinklers in the work zone.
7 Fire alarm system to be put on test mode during the drain down to ensure other monitoring modules in the fire sprinkler system are not activated as part of the drain down process.
8 Fire hoses to be connected to the drain connection and extended to the permitted drain area.

9 Once the system is drained down, the system is to be put back online with the exception of the zone being worked on.
10 Contractor to perform scheduled work.
11 FPPM, contractor and fire alarm service vendor to meet at the end of impairment time to discuss work status.
12 When work is complete, the fire alarm service vendor is to place the fire alarm system back in test mode to allow for the refilling of the system.
13 If a fire pump is provided in the building, power to be shut off to fire pump ahead of refilling the system to prevent the activation of the pump.
14 Sprinkler contractor to refill the system slowly to prevent any air pockets or hammering. Contractor to confirm there are no leaks in the modified system.
15 Once filled, the FPPM shall verify that the control valve is in the fully open position.
16 Fire alarm service vendor to reenable disabled sprinkler module points.

Fire Alarm Impairments:

[No fire alarm system impairments are anticipated as part of the project.]

[It is anticipated that impairments to the fire alarm system will be necessary as part of the project. Due to the installation of {insert specific elements such as ceiling assemblies, partitions, etc.}, modifications to the fire alarm system are necessary to maintain adequate fire alarm coverage. {Describe sequencing approaches that address how the fire alarm system will be maintained throughout the various stages of construction. For example, describe how devices will need to be relocated as new partitions and other obstructions are installed.}

{Specific fire alarm impairment procedures should be outlined here}. Fire alarm impairments must be coordinated with the owner's FPPM by the submission of a fire protection impairment request form. Fire alarm impairments must be carried out in accordance with the following protocols:

1 Analysis of work tasks to determine the risk of activation of the smoke detector or the need for other fire alarm device impairment.
2 Impairment request form submitted from contractor performing work to project FPPM.
3 Project FPPM to review requests against other work occurring, extent of impairment and duration.
4 If approved, project FPPM to schedule fire alarm service vendor to disable requested devices for the identified duration.
5 Project FPPM to notify the project team, including property managers, building engineers, superintendents, of scheduled impairment to ensure the greater team is aware.

6 FPPM, fire alarm service vendor, building engineers and contractor to meet at the fire command center (if provided), or pre-determined location in close proximity to the fire alarm control panel at the beginning of impairment start time.
7 FPPM to verify that the correct/required devices are disabled within the work zone, working with the fire alarm service company and contractor.
8 Protective covers or "bags" to be placed on impaired smoke detectors to protect devices from damage.
9 Contractor to perform scheduled work.
10 FPPM, contractor and fire alarm service vendor to meet at the end of impairment time to discuss work status.
11 FPPM to verify with the contractor that dust-generating work is complete and the work environment is not dusty, or relocation of the fire alarm device is complete.
12 When work is complete, protective covers or "bags" are to be removed from impaired devices. FPPM to verify protective covers have been removed from all devices. *This step must occur prior to step 13 to prevent accidental activation by creating a suction when removing the cover.
13 Fire alarm service vendor to re-enable impaired smoke detection devices.

Smoke Control System Impairments:

[No smoke control system impairments are anticipated as part of the project.]

[It is anticipated that impairments to the smoke control system will be necessary as part of the project. Due to the installation of {insert specific elements such as ceiling assemblies, partitions, etc.}, modifications to the smoke control system are necessary. {Describe sequencing approaches that address how the smoke control system will be maintained throughout the various stages of construction.}

{Specific smoke control impairment procedures should be outlined here}. Smoke control impairments must be coordinated with the owner's FPPM by the submission of a fire protection impairment request form. Smoke control system impairments must be carried out in accordance with the following protocols:

{Specific impairment procedures must be developed to address the various types of smoke control systems present and interconnectivity to other fire protection systems. It is recommended that a licensed engineer be engaged to review and propose impairment procedures, as the complexity surrounding smoke control system impairments is much more extensive than fire alarm or fire sprinkler.}

6. Hot Work Procedures:

[Hot work is not anticipated as part of the project.]

[Hot work is anticipated as part of the project. It is expected that the following activities will require hot work {insert activity and type of hot work. For example, welding steel members for the installation of a new stair.}.

Hot work is defined as {insert definition of hot work based on location of project and applicable codes and standards}.

The following protocols should be implemented for all hot work operations {this section should be updated to align with applicable codes, standards and permitting requirements based on project location.} (Table B.2)

1 A "Hot Work Permission Request from Owner" form (Form 12.A) must be submitted to the Owner for review and approval for each subcontractor who will be performing hot work.
2 Requests will be reviewed and approved at the sole discretion of the Owner's FPPM. If approved, the FPPM will grant an Owner's Hot Work Permission Letter to the general contractor/subcontractor.
3 Subcontractor to apply for a hot work permit from the Fire Department using the permission letter provided.
4 Once a hot work permit from the fire department has been granted, a copy must be provided to the Owner's FPPM and General Contractor's FPPM.

In addition to the city/town permitting requirements, the following internal project hot work permitting system will be implemented by the general contractor:

1 A daily hot work permit must be filled out by each subcontractor performing hot work. The permit must be posted at the designated project command center. The permit must include the date and time, name of contractor and worker conducting the hot work and their certificate number, the location and must be signed by the general contractor's FPPM prior to any daily hot work.
2 A daily hot work log (Table 12.1) and a fire watch log must be completed and up to date at the designated command center. Records must be maintained of these logs for the duration of the project.
3 A fire watch must be arranged at all hot work operations. A fire watch detail from the fire department is required if noted on the Hot Work Permit.
4 A Fire extinguisher must be provided at every hot work location suitable for the possible type of fire.
5 All combustible items must be removed from the work area and any area where byproducts of the hot work could land. Therefore, special precautions must be taken to protect floor openings and the edge of building areas where sparks or slag could fall. NFPA 51B, *Standard for Fire Prevention During Welding, Cutting, and Other Hot Work*, recommends removing all combustible materials within 35 feet of the hot work. Where this is not feasible, fire blankets or other noncombustible shields or barriers should be used to protect combustible items and to prevent sparks from traveling far from the hot work site.
6 Ensure sprinkler system is not impaired within the same zone as hot work activities. It is critical that the sprinkler system remains active; otherwise, the building is put in an extremely vulnerable situation.

7 Evaluate work zone for fire alarm devices. Smoke detectors in close proximity to hot work operations should be disabled and covered, or "bagged," to protect the devices from dust and other debris. Impairment procedures established in CFSP are to be followed.

Safeguards During Hot Work Operations

1 Adequate Personal Protective Equipment (PPE) must be worn by all personnel involved with the hot work operation.
2 Ensure combustibles remain outside of the work zone.
3 Ensure protection deployed stays in place during the course of work.
4 The fire watch must be on-site for the entire duration of the hot work.

Closeout of Hot Work Activities:

1 The fire watch must remain on-site for the specified time after the completion of work in compliance with applicable codes and standards. {Typically, 30-minute, 1-hour, 2-hour time frames are specified}. This is critical as hot spots can ignite and start a fire even after the work is completed. Applicable NFPA codes such as NFPA 1, NFPA 51B and NFPA 241 will govern this. However, other requirements should be reviewed, such as local regulations, insurance requirements and other building-specific criteria.
2 Hot work logs must be updated to indicate end times.
3 Impaired fire alarm devices must be brought back into service and verified by the FPPM.

7. Waste Management Plan:
[Combustible debris and other byproducts of construction will be collected, organized and removed from the jobsite at the end of each work shift. {Insert description of waste removal plan, such as the use of trash chute, dumpsters, live-loading removal, etc.}]

8. Inspections:
[The FPPM is responsible for conducting periodic inspections of the construction site to ensure compliance with the construction fire safety plan. It is the responsibility of the FPPM to maintain records of the inspections for the duration of the project, as they may be requested by the AHJ, ownership or insurance agencies.]

Include a tailored inspection checklist that includes project-specific elements. (Chapter 15 provides an example of the contents of the checklist).

9. Pre-Fire Plan:
[The Pre-Fire Plan shows the location of pertinent fire protection and life safety elements at the building.] The Pre-Fire Plan identifies the location of the following elements: (Figure B.1)

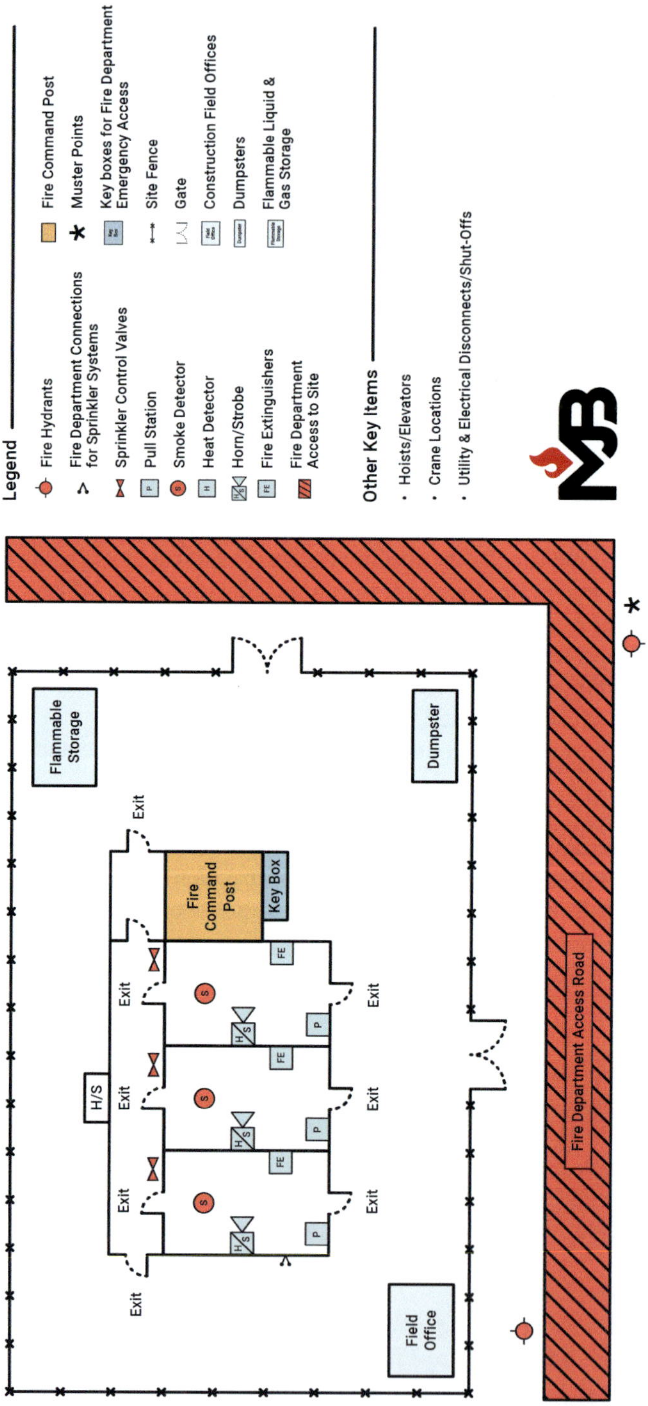

Figure B.1 An example of a pre-fire plan and key components.

Exterior elements:

1 Site fence and gate access
2 Fire hydrants
3 Fire department access
4 Fire department key boxes
5 Site entrances
6 Fuel storage areas
7 Personal transportation vehicles
8 Cooking locations

Interior elements:

1 Means of egress
2 Fire alarm system components
3 Fire sprinkler system components
4 Fire extinguishers
5 Fire command post

Index

Note: Page references with *Italics* refer to figures, **bold** refer to tables.

For Product Safety Concerns and Information please contact our
EU representative GPSR@taylorandfrancis.com Taylor & Francis
Verlag GmbH, Kaufingerstraße 24, 80331 München, Germany